Lanthanides and actinides

Simon Cotton
Felixstowe College

Oxford University Press
New York 1991

Published in Great Britain by Macmillan Education Ltd

Published in the United States by
Oxford University Press, Inc
200 Madison Avenue, New York 10016

Library of Congress Cataloging-in-Publication Data
Cotton, Simon.
Lanthanides and actinides / Simon Cotton.
p. cm.
Includes bibliograhical references and index.
ISBN 0-19-507366-5
1. Rare earth metals. 2. Actinide elements. I. Title.
QD172.R2C67 1991
546′.4—dc20
91-17815
CIP

Printing (last digit) 9 8 7 6 5 4 3 2 1

Printed in Hong Kong

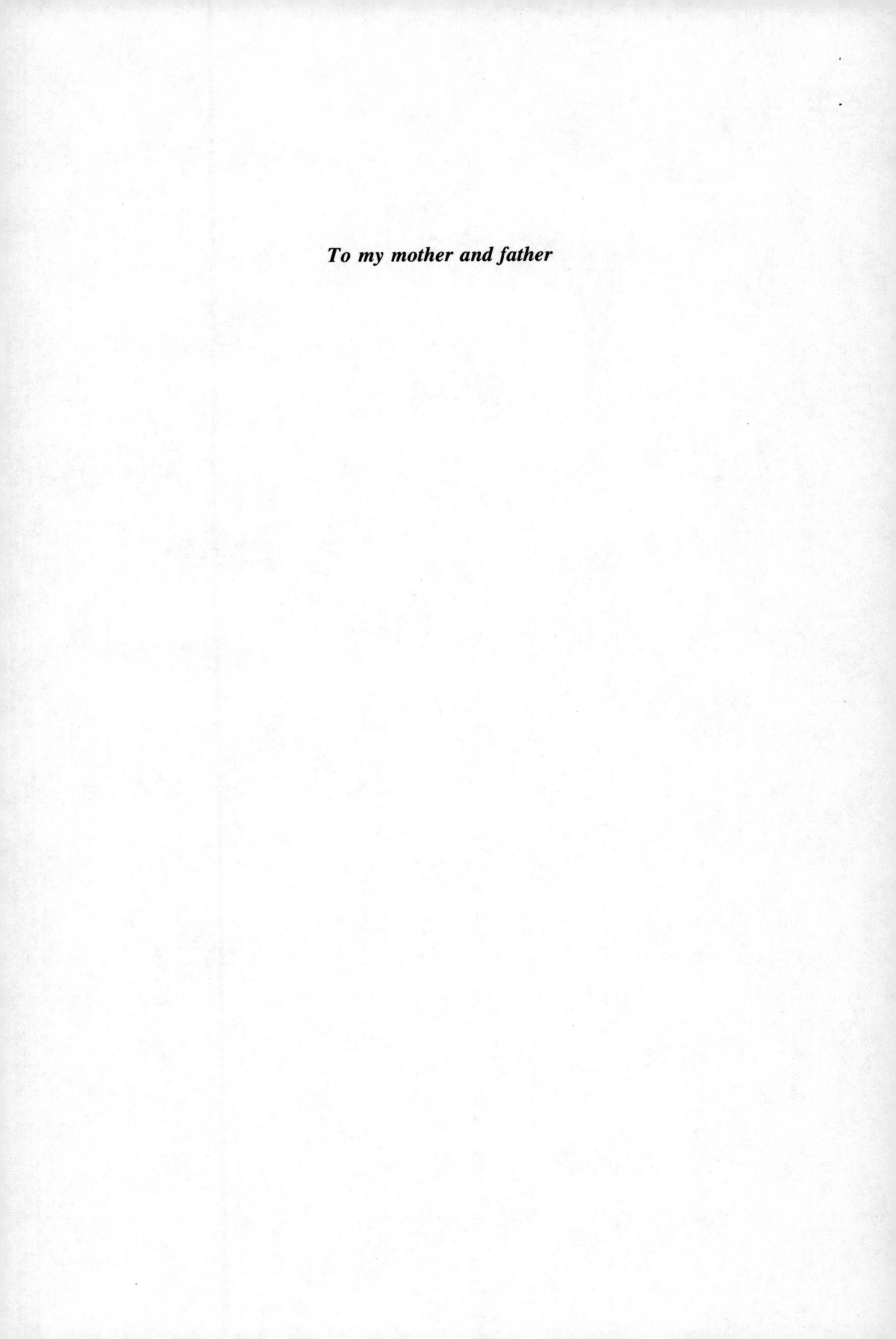

To my mother and father

Contents

Preface

This book is intended to provide an adequate background to the chemistry of the *f*-block metals for students on advanced undergraduate courses, as well as for postgraduates. It is hoped that it will also be of value to teachers requiring an up-to-date account of the descriptive chemistry of these metals.

All books reflect the interests (and prejudices) of their authors; this one is no exception in that it concentrates its efforts in attempting to give a sound factual foundation, relying on the Bibliography (at the end of the book) to give the reader access to a wider spectrum of the literature.

It is a very real pleasure to thank many people for their contributions to this book:

Drs D M Adams and M Green, for helpful comments on the manuscript; members of the secretarial staff of Stanground College – Jane, Sue, Anne, and all 'the girls' for their help with the production of parts of the manuscript, especially some awkward tables; Kerry Lawrence and her colleagues at Macmillan Education for their tolerance; and the Chemistry Department of Cambridge University for library facilities.

I would wish to single out Dr Alan Hart, without whose collaboration, advice and influence I would never have been in a position to undertake this book; Dom Philibert Zobel and the monastic community of Bec for the hospitality of their roof while I compiled the Index and Bibliography; and last, but by no means least, Carolyn Wright who uncomplainingly translated my writing into typescript.

Simon Cotton
Abbaye N. D. du Bec
Feast of the Transfiguration
1990

Acknowledgements

The author and publishers wish to thank the following who have kindly given permission for the use of copyright material:

The American Chemical Society for Figures 1.1, 2.13, 2.14, 2.27, 3.15 and 3.41;
The Editor of the *Journal of Chemical Education* for Figure 2.2;
Dr S. Hüfner and Kluwer Academic Publishers for Figure 2.9;
Pergamon Press PLC for Figures 2.14(b), 3.7 and 3.13;
The Royal Society of Chemistry for Figures 2.17, 2.21, 2.33, 2.38 and 3.28;
Elsevier Scientific Publications for Figure 3.9;
The Editor of *Inorganica Chimica Acta* for Figures 3.14, 3.16 and 3.17.

Every effort has been made to trace all the copyright holders but if any have been inadvertently overlooked the publisher will be pleased to make the necessary arrangement at the first opportunity.

1

Scandium

1.1 Introduction

Scandium is probably the least-studied element in the fourth period. It is commonly grouped with yttrium and the lanthanides in Group 3 (IIIA), yet its atom (and ion) are considerably smaller than those of yttrium and lanthanum; it is however larger than Al^{3+} (with which it is often compared; ionic radius 0.5 Å in Group 13 (IIIB)). Table 1.1 compares the coordination numbers of scandium and lanthanides in a selection of coordination compounds with those of iron(III), a d^5 ion with no ligand-field effects dictating the adoption of a particular coordination number.

Despite a chemistry limited to one oxidation state (III) and restrictions on grounds of cost (1991 price, about £100 per g), interest in the element lies in its adoption of a wide range of coordination numbers while there is a small but burgeoning selection of organometallics.

Scandium ores

Present at a level of some 10–20 ppm, scandium is relatively evenly distributed, and this partly accounts for its cost – compare with U, Hg, Ag and Au (4, 0.5, 0.1 and 0.0005 ppm respectively).

The three natural scandium minerals are thortveitite, sterrite and kolkbeckite. The former is a silicate, $Sc_2Si_2O_7$, found as grey-green crystals along with various lanthanide ores, principally in the Malagasy Republic and Norway. Sterrite ($ScPO_4.2H_2O$) and the related kolbeckite are rarer. Several minerals containing the lanthanides and uranium also contain scandium at a low level; much scandium is obtained from davidite (0.02 per cent Sc) as a by-product from uranium manufacture, a result of the widespread occurrence of the ore.

Table 1.1 Comparison of coordination numbers in complexes of Sc^{3+}, Fe^{3+} and Ln^{3+}

	Sc	Fe	Ln^{3+}
Radius M^{3+} (Å)	0.745	0.64	1.03 (La^{3+})
			0.85 (Lu^{3+})
Ligand			
H_2O	6?	6	9
	$Sc(OH_2)_6^{3+}$	*$Fe(OH_2)_6^{3+}$	*$Nd(OH_2)_9^{3+}$
$(CH_3COCHCOCH_3)^-$	6	6	8
acac	*$Sc(acac)_3$	*$Fe(acac)_3$	*$La(acac)_3(H_2O)_2$
NO_3^-	9	8	10–12
	*$Sc(NO_3)_5^{2-}$	*$Fe(NO_3)_4^-$	*$Ce(NO_3)_6^{3-}$, $La(NO_3)_5(OH_2)_2^{2-}$
EDTA	7	7	9
	$Sc(EDTA)(OH_2)^-$	*$Fe(EDTA)(OH_2)^-$	*$La(EDTA)(OH_2)_3^-$
F^-	6	6	9
	*ScF_6^{3-}	*FeF_6^{3-}	*LaF_3
Cl^-	6	4, 6	6, 8, 9
	*$ScCl_3$	*$FeCl_4^-$, $FeCl_3$	*$LnCl_3$ (6 for Tb–Lu)
			(9 for La–Gd)
		$FeCl_6^{3-}$	(8 for Tb)
$N(SiMe_3)_2^-$	3	3	3, 4, 5
(btsa)	*$Scbtsa_3$	*$Febtsa_3$	*$Lnbtsa_3$
			*$Lnbtsa_3$, Ph_3PO
			*$Lnbtsa_3(Me_3PO)$
			$Lnbtsa_3(Me_3PO)_2$
2,2′,2″-terpyridyl	(8 + 1)	6	9
(terpy)	*$Sc(terpy)(NO_3)_3$	Fe terpy Cl_3	*$Eu(terpy)_3^{3+}$
		$Fe(terpy)_2^{3+}$	*$Pr(terpy)Cl(OH_2)_5^{2+}$
Pyridine *N*-oxide	6	6	8
(pyO)	$Sc(pyO)_6(ClO_4)_3$	$Fe(pyO)_6(ClO_4)_3$	*$La(pyO)_8(ClO_4)_3$
Ph_3PO	8	8?	10
	*$Sc(Ph_3PO)_2(NO_3)_3$	$Fe(Ph_3PO)_2(NO_3)_3$	*$La(Ph_3PO)_4(NO_3)_3$

* = confirmed by X-ray diffraction.

Extraction of the element from its ores first involves digestion. A silicate like thortveitite yields the soluble chloride on digestion with HCl, whereas HF treatment affords the insoluble fluoride, subsequently dissolved by sulphuric acid. Separation from the lanthanides is possible by ion-exchange methods. The soft silvery metal (mp 1541°) is made by calcium reduction of ScF_3 at 1600°; it is readily dissolved by acid.

1.2 Binary compounds of scandium

These are formed with a variety of non-metals. Scandium oxide, Sc_2O_3, a white solid (mp 3100°) is made by ignition of the metal or compounds such as the nitrate or hydroxide. It is basic and has a body-centred cubic structure with 6-coordinate scandium. It dissolves in excess NaOH to afford a species which may be $Sc(OH)_6^{3-}$.

The halides can be obtained by various methods:

$$Sc \xrightarrow[225°]{HF_2} ScF_3 \qquad \text{(white solid)}$$

$$ScCl_3.6H_2O \xrightarrow[\text{reflux}]{SOCl_2} ScCl_3 \qquad \text{(white solid mp 939°)}$$

$$ScBr_3.nH_2O \xrightarrow[\text{heat}]{P_2O_5} ScBr_3 \qquad \text{(white solid mp 948°)}$$

$$Sc \xrightarrow[600°]{I_2} ScI_3 \qquad \text{(yellow solid mp 920°)}$$

$$(NH_4)_3ScX_6 \xrightarrow{500°} ScX_3 + 3NH_4X \ (X = Cl,Br)$$

ScF_3 has octahedral coordination of scandium in the WO_3 structure; like the lanthanide fluorides, it is insoluble in water but dissolves in HF or NH_4F as complex ions such as ScF_6^{3-}. Anhydrous scandium chloride, a very hygroscopic solid, has the $FeCl_3$ structure with 6-coordinate scandium; unlike $AlCl_3$ it is not a Friedel–Crafts catalyst, the larger radius of scandium presumably contributing to a lower polarising power and Lewis acidity. $ScBr_3$ and ScI_3 also have the $FeCl_3$ structure (also referred to as the BiI_3 structure). The chloride and the bromide can both be sublimed *in vacuo* at 800–1000°. Similar octahedral coordination is found in halide complexes such as Na_3ScF_6 (cryolite structure), $(NH_4)_3ScF_6$ and $Cs_2NaScCl_6$ (fcc). $(NH_4)_3ScF_6$ exists in high (α) and low-temperature (β) forms: ν Sc–F has been assigned to a band at 504 cm^{-1} in its Raman spectrum. Other complexes, presumably with bridges, like $NaScF_4$ have been made.

Lower halides

These contain scandium with a formal oxidation state less than three. They are generally made by reaction of Sc with $ScCl_3$ at high temperature:

$$Sc + ScCl_3 \xrightarrow{950^\circ} Sc_5Cl_8$$

$$Sc + ScCl_3 \xrightarrow{890^\circ} Sc_7Cl_{12}$$

ScCl and Sc_7Cl_{10} have also been made. Several structures are known (Figure 1.1) and in general there is some metal–metal bonding. Sc_7Cl_{12} can be regarded as $Sc^{3+}(Sc_6Cl_{12})^{3-}$ with anions similar to the well-known M_6Cl_{12} clusters formed by Nb and Ta; the Sc–Sc distance in the anions is 3.20–3.23 Å (3.26 Å in scandium metal).

Other binary compounds generally formed by direct reaction include ScC (cubic), ScB (mp 2250°) and ScH_2 (fcc). The hydride is highly conducting and may have a structure of the type $Sc^{3+}(H^-)_2(e^-)$.

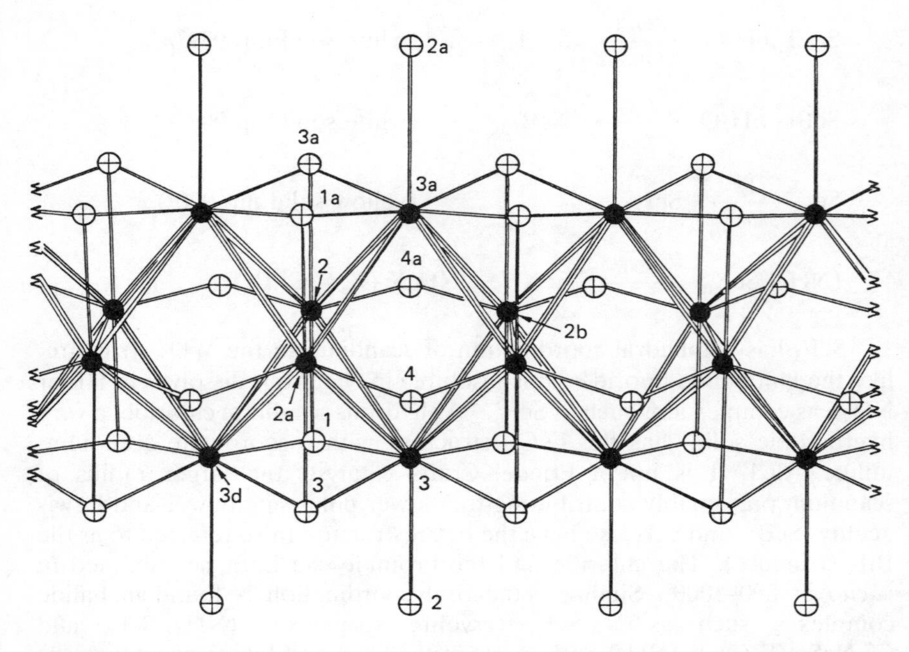

Figure 1.1 The anionic polymeric chain in Sc_5Cl_8, showing the edge-sharing octahedra (solid ellipsoids represent scandium, open ellipsoids chlorine) [reprinted with permission from K. R. Poppelmeier and J. D. Corbett, *J. Amer. Chem. Soc.*, **100** (1978) 5042, copyright 1978 American Chemical Society]

Simple salts

No definite data are available on the aquo-ion which is probably $[Sc(OH_2)_6]^{3+}$; in aqueous solution it complexes with chloride, nitrate, and even perchlorate.

The nitrate and sulphate are both obtained as colourless crystals of $Sc(NO_3)_3.4H_2O$ and $Sc_2(SO_4)_3.5H_2O$ respectively. They lose water at above 100° and afford the oxide at 250° (nitrate) and 1000° (sulphate). Other salts include the oxalate, $Sc_2(C_2O_4)_3.xH_2O$, obtained by precipitation; unlike the lanthanide oxalates it is slightly soluble (about 60 mg l^{-1}) and dissolves readily in excess oxalate (as $Sc(C_2O_4)_2^-$). No normal carbonate is known, basic carbonates being precipitated.

1.3 Complexes of scandium

1.3.1 Oxygen-donor ligands

A considerable number of complexes of monodentate ligands such as THF, pyridine *N*-oxide, triphenyl phosphine oxide, hexamethylphosporamide and dimethylsulphoxide have been prepared. They fall broadly into two series, ScL_6X_3 (where X is a non-coordinating anion such as perchlorate) and ScL_3X_3 (where X = Cl, NCS). X-ray data are lacking except for mer-$ScCl_3$ $(THF)_3$, but six-coordination is probable for the others too. The nitrate group with its smaller 'bite' angle permits the higher coordination number of 8 to be reached in $[Sc\{(NH_2)_2CO\}_4(NO_3)_2]^+NO_3^-$ and $Sc(Ph_3PO)_2(NO_3)_3$ (both X-ray).

Chelating ligands such as β-diketones form many complexes with scandium. $Sc(acac)_3$ has been shown to be six-coordinate (X-ray) as doubtless are the other tris-chelates, with ligands like dipivaloylmethane. This is in contrast to the behaviour of the lanthanides which, with all except the most bulky ligands, form either dimers or hydrated monomers with 7 or 8-coordination. Some tetrakis complexes like $M^+Sc\ (CF_3COCHCOCF_3)_4^-$ (M = alkali metal) are most likely 8-coordinate and sublime *in vacuo*. Similar complexes with the tropolonate ion ($Sc(trop)_3$ and $NaSc(trop)_4$) have also been characterised and have 6- and 8-coordinate structures respectively. Although the structure of scandium nitrate is unknown, the complex $Sc(NO_3)_3.N_2O_4$ has been shown to be $(NO^+)_2[Sc(NO_3)_5]^{2-}$ with 9-coordinate scandium, one nitrate being monodentate. High coordination numbers are also possible for crown ether complexes, but no X-ray data are yet known for complexes where Sc^{3+} is bound to the ether. In contrast, the bulky alkoxide ligand 2, 6-di-t-butyl-4-methylphenate (OR) affords the 3-coordinate $Sc(OR)_3$ (X-ray) and adducts $Sc(OR)_3.L$ (L = Ph_3PO,THF) which are probably 4-coordinate.

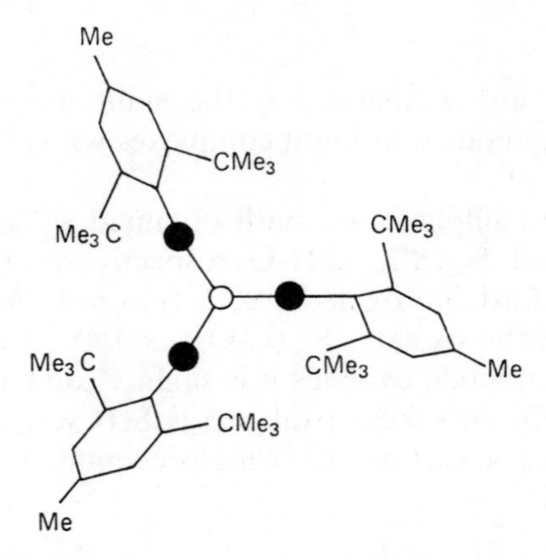

Figure 1.2 The three coordinate scandium compound scandium tris(2,6-di-t-butyl-4-methylphenoxide)

1.3.2 Nitrogen-donor ligands

Reaction of pyridine with the anhydrous halides affords $Scpy_4X_3$ (X = Cl, Br) which, like the iron analogues, are probably $Scpy_3X_3.py$; *in vacuo*, $Scpy_3X_3$ are formed. More complexes have been obtained (using suitable non-aqueous solvents such as THF) with 1,10-phenanthroline, ethylenediamine, and 2,2-bipyridyl and 2,2′:6′,2″-terpyridyl. Of these $Sc(terpy)(NO_3)_3$ has 9-coordinate scandium, but with one very long Sc–O bond.

Macrocyclic ligands such as tetraphenylporphyrin (TPP), octaethylporphyrin (OEP) and phthalocyanine afford both monomeric complexes like Sc(OEP)(OAc) and dimeric (OEP)ScOSc(OEP) similar to those of other trivalent metal ions.

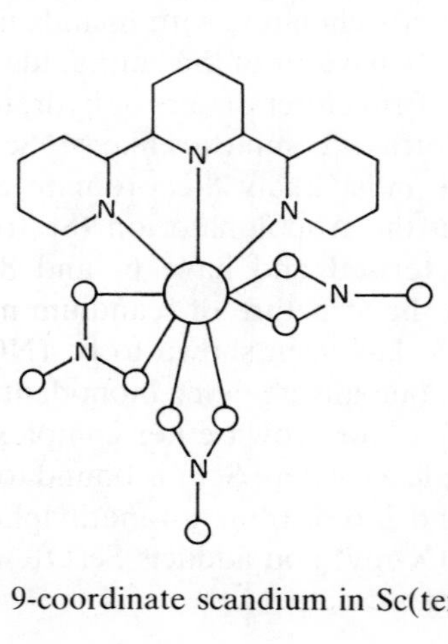

Figure 1.3 9-coordinate scandium in $Sc(terpy)(NO_3)_3$

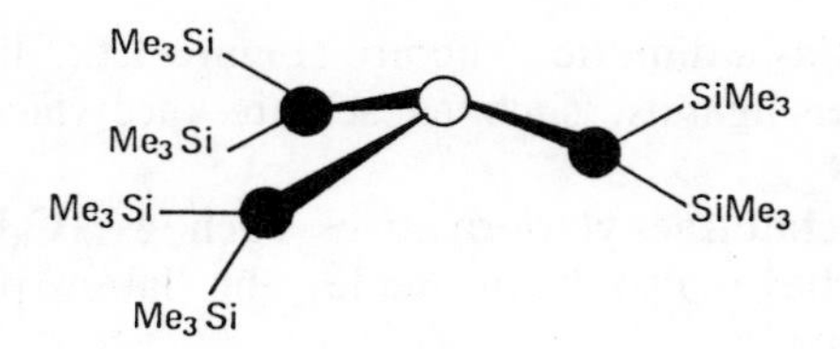

Figure 1.4 The pyramidal $Sc[N(SiMe_3)_2]_3$

Like the early 3*d* metals, uranium and the lanthanides – scandium forms a 3-coordinate silylamide. It is formed from the reaction of $ScCl_3$ with $LiN(SiMe_3)_2$ in THF. $Sc[N(SiMe_3)_2]_3$ resembles the lanthanide analogues in having a trigonal pyramidal structure – the 3*d* tris-silylamides being planar. It is readily soluble in such covalent solvents as benzene and pentane. Unlike the corresponding lanthanide compounds (section 2.10.3), it shows no tendency to increase its coordination sphere by adduct formation.

1.4 Organometallic compounds of scandium

The first compounds to be made were simple cyclopentadienyls. Thus reacting scandium chloride with excess C_5H_5Na affords $Sc(C_5H_5)_3$ as a straw-coloured solid, soluble in organic solvents, thermally stable (mp 240°) and subliming *in vacuo*; like other scandium organometallics, it is immediately hydrolysed by water. It has a polymeric structure in which each scandium is bound η^5 to two cyclopentadienyls and η^1 to two others, in contrast to lanthanide cyclopentadienyls (typically $3\eta^5$ and $1\eta^1$), a consequence of the smaller ionic radius of scandium.

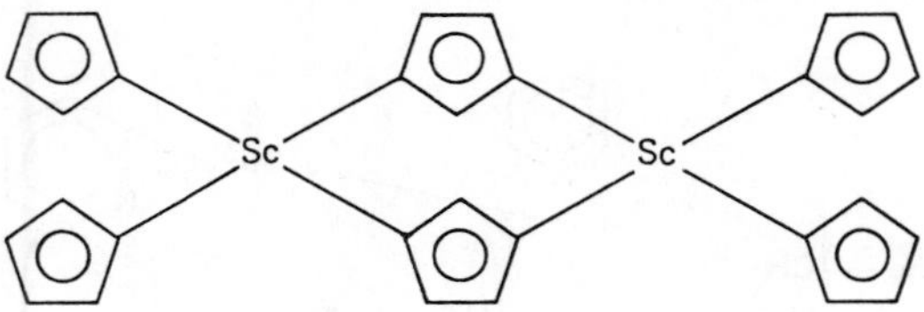

Sc Sc

Figure 1.5

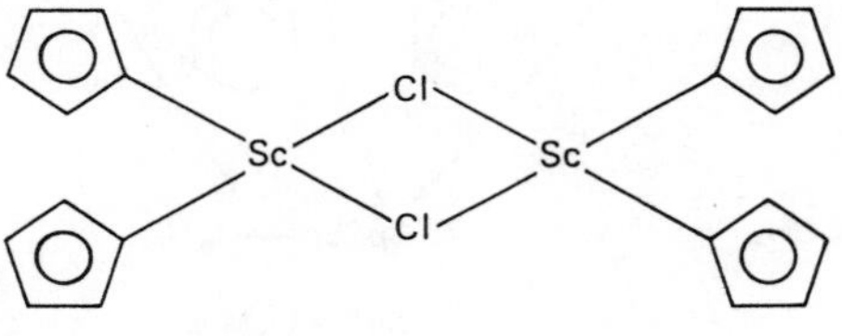

Cl Sc Sc Cl

Figure 1.6

$Sc(C_5H_5)_2Cl$ has a dimeric structure (Figure 1.6). The chlorine can be replaced by other ligands, such as acetate, acetylacetonate and alkyl groups.

Some cyclooctatetraenyl derivatives such as $(C_8H_8)ScCl.THF$ and $(C_8H_8)Sc(C_5H_5)$ have also been made, the latter probably having a 'sandwich' structure.

The most interesting organoscandium compounds of this type have been reported for the pentamethylcyclopentadienyl ligand, C_5Me_5. Its bulk means that only two C_5Me_5 groups can bind to scandium, so that the monomeric $(C_5Me_5)_2ScCl$ (which forms a labile THF adduct) is isolated. This may be converted into alkyls $(C_5Me_5)_2ScR$ ($R = CH_3$, C_6H_5, $C_6H_5CH_2$) which are ethene polymerisation catalysts and react with hydrogen to form an unstable hydride $(C_5Me_5)_2ScH$ (which in turn undergoes alkene insertion reaction to form alkyls). The hydride catalyses H/D exchange between H_2, arenes, and the primary C–H bonds of molecules like PMe_3 and $SiMe_4$. In toluene, the methyl hydrogens exchange about 60 times more slowly than the ring hydrogens. Some of the reactions are shown in Figure 1.7; the metallation reaction with pyridine to afford a compound where scandium is bound both to nitrogen and an adjacent carbon atom is most interesting.

σ-bonded alkyls and aryls

The first authentic compound of this type, reported in 1968, was triphenylscandium, a pyrophoric yellow-brown solid. Shortly afterwards the alkyls, $Sc(CH_2QMe_3)$ ($Q = C$, Si) and $Sc[CH(SiMe_3)_2]_3.2THF$ were made, these being air-sensitive white crystalline solids. More recently, the $ScMe_6^{3-}$ ion, apparently octahedral, has been obtained as the very air and moisture sensitive $Li(Me_2N(CH_2)_2NMe_2)^+$ salt.

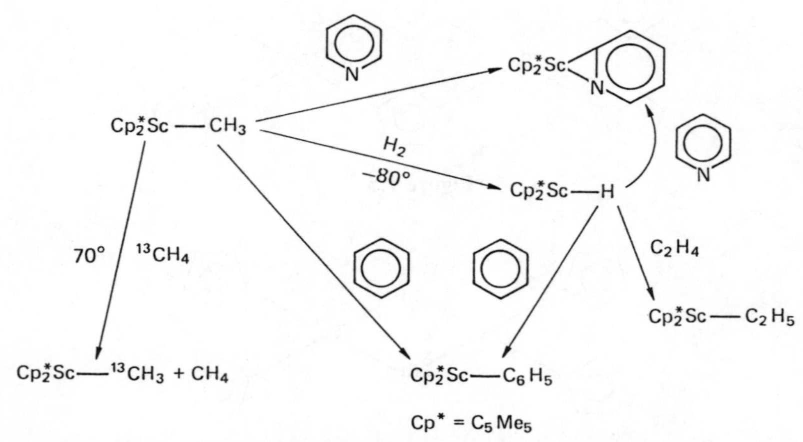

Figure 1.7 Reactions of pentamethylcyclopentadienyl compounds

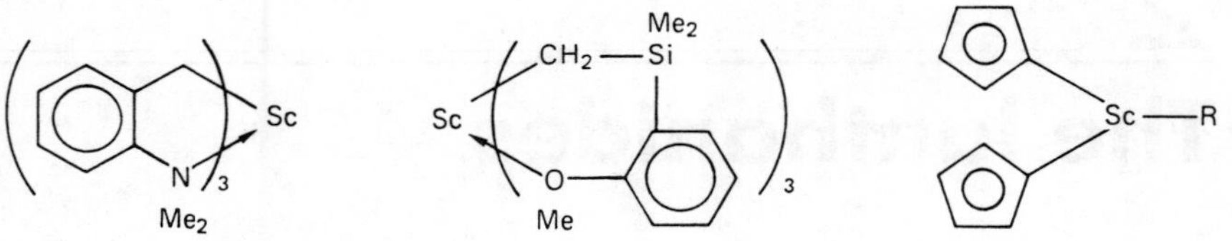

Me2
CH2—Si
Sc
Sc
N
3
Me2
O
Me
3
Sc—R

I I II

Figure 1.8

A number of the other complexes have been isolated in which the ligands are not simple aryls or alkyls. One type involves chelating ligands (I); the second kind have 'supporting' cyclopentadienyl ligands (II) (see Figure 1.8).

Reaction of $(C_5H_5)_2ScCl$ with $LiAlMe_4$ affords the pale yellow $(C_5H_5)_2ScMe_2AlMe_2$, sublimeable *in vacuo* at 100°. This is stereochemically rigid on the NMR timescale up to 100°, in contrast to the yttrium analogue, fluxional at 40°; the increased readiness of yttrium to exchange bridging and terminal methyls is presumably related to steric congestion around scandium. On cleavage with Lewis bases, the complexes $(C_5H_5)_2ScMe.L$ (L = THF, py) are formed – unlike the lanthanides, which form dimeric $(C_5H_5)_2LnMe_2Ln(C_5H_5)_2$ (see page 78).

2 The lanthanides

2.1 Position of the lanthanides in the Periodic Table

In 1794, J. Gadolin obtained yttria, impure yttrium oxide, from the mineral now known as gadolinite, while in 1803, Berzelius and Klaproth isolated the first cerium compound. Subsequent research over the ensuing century demonstrated, first, that yttria was a mixture of the oxides of yttrium, erbium and terbium; and that cerite was made up of the oxides of lanthanum and cerium (Mosander, 1839–43); secondly that the original 'erbium oxide' in fact contained erbium, holmium, thulium and ytterbium (Cleve, Marignac, 1878–80). Similarly, Mosander's separation of 'didymium' from lanthanum was later eclipsed by its separation into samarium, europium, neodymium and praseodymium.

The difficulties in separating the lanthanides and obtaining the elements in a pure form, a problem not entirely resolved until after the Second World War, were basically due to the pronounced chemical similarities between the elements. By 1907, with the discovery of lutetium, all the elements had been identified, save the radioactive promethium (section 2.13); Moseley used X-ray spectra of the elements to place them in the Periodic Table (1913), showing that there were 14 elements between lanthanum and hafnium, even leaving a gap for the yet-to-be-discovered element with atomic number 61.

As the lanthanide series is traversed from lanthanum to lutetium, the atomic (and ionic) radii decrease. The 4*f* electrons are 'inner' electrons in the sense that the maximum of their charge density functions is well inside the outermost ($5s^2 5p^6$) electrons; thus they are shielded from the surroundings of the lanthanide ion (accounting for the limited dependence of *f–f* spectra of complexes on ligand). The 5*s* and 5*p* orbitals, however, penetrate the 4*f* subshell and thus are not shielded from the increasing nuclear change, contracting as the lanthanide series is traversed.

Table 2.1 Some properties of lanthanide ions

	EC of metal	EC of Ln^{3+}	Colour of Ln^{3+}	$(^{2S+1}L_J)$ Ground state	μ_{eff} (Hund)	μ_{eff} (Van Vleck)	$M(phen)_2(NO_3)_3$	μ_{eff} $M(C_5H_5)_3$	$M_2(SO_4)_3 \cdot 8H_2O$	$M(dpm)_3$
Lanthanum	$(Xe)5d^16s^2$	$(Xe)4f^0$	Colourless	$^1S_0$	0	0	0	0	0	0
Cerium	$4f^15d^16s^2$	f^1	Colourless	$^2F_{5/2}$	2.54	2.56	2.46	2.46	2.39	—
Praesodymium	f^3s^2	f^2	Green	$^3H_4$	3.58	3.62	3.48	3.61	3.62	3.65
Neodymium	f^4s^2	f^3	Lilac-red	$^4I_{9/2}$	3.62	3.68	3.44	3.63	3.62	3.6
Promethium	f^5s^2	f^4	Pink	$^5I_4$	2.68	2.83	—	—	—	—
Samarium	f^6s^2	f^5	Pale yellow	$^6H_{5/2}$	0.845	1.55–1.65	1.64	1.54	1.54	2.05
Europium	f^7s^2	f^6	Colourless	$^7F_0$	0	3.4–3.5	3.36	3.74	3.61	3.5
Gadolinium	$f^7d^1s^2$	f^7	Colourless	$^8S_{7/2}$	7.94	7.94	7.97	7.98	7.95	7.7
Terbium	f^9s^2	f^8	Very pale pink	$^7F_6$	9.72	9.7	9.81	8.9	9.6	9.6
Dysprosium	$f^{10}s^2$	f^9	Pale yellow	$^6H_{15/2}$	10.60	10.6	10.6	10.0	10.5	10.3
Holmium	$f^{11}s^2$	f^{10}	Yellow	$^5I_8$	10.61	10.6	10.7	10.2	10.5	10.0
Erbium	$f^{12}s^2$	f^{11}	Rose pink	$^4I_{15/2}$	9.58	9.6	9.46	9.45	9.55	9.3
Thulium	$f^{13}s^2$	f^{12}	Pale green	$^3H_6$	7.56	7.6	7.51	7.1	7.2	7.2
Ytterbium	$f^{14}s^2$	f^{13}	Colourless	$^2F_{7/2}$	4.54	4.54	4.47	4.0	4.4	4.3
Lutetium	$f^{14}d^1s^2$	f^{14}	Colourless	$^1S_0$	0	0	0	0	0	0
Yttrium	$(Kr)4d^15s^2$	(Kr)	Colourless	$^1S_0$	0	0	0	0	0	0

There is general agreement from spectroscopic data on the electronic structure of most lanthanide atoms (Table 2.1). Right at the beginning of the series, with lanthanum, the 5*d* subshell is lower in energy than 4*f*, so that the ground-state configuration of La is $[Xe]5d^1 6s^2$. Opinions differ over whether Ce is $[Xe]4f^1 5d^1 6s^2$ or $[Xe]4f^2 6s^2$; at all accounts, 4*f* and 5*d* are of comparable energy, and as additional protons are added to the nucleus, the 4*f* orbitals rapidly contract and are stabilised relative to 5*d* (owing to their greater penetration of the xenon core) so that praesodymium has the ground state $[Xe]4f^3 6s^2$, and this trend continues as far as europium,

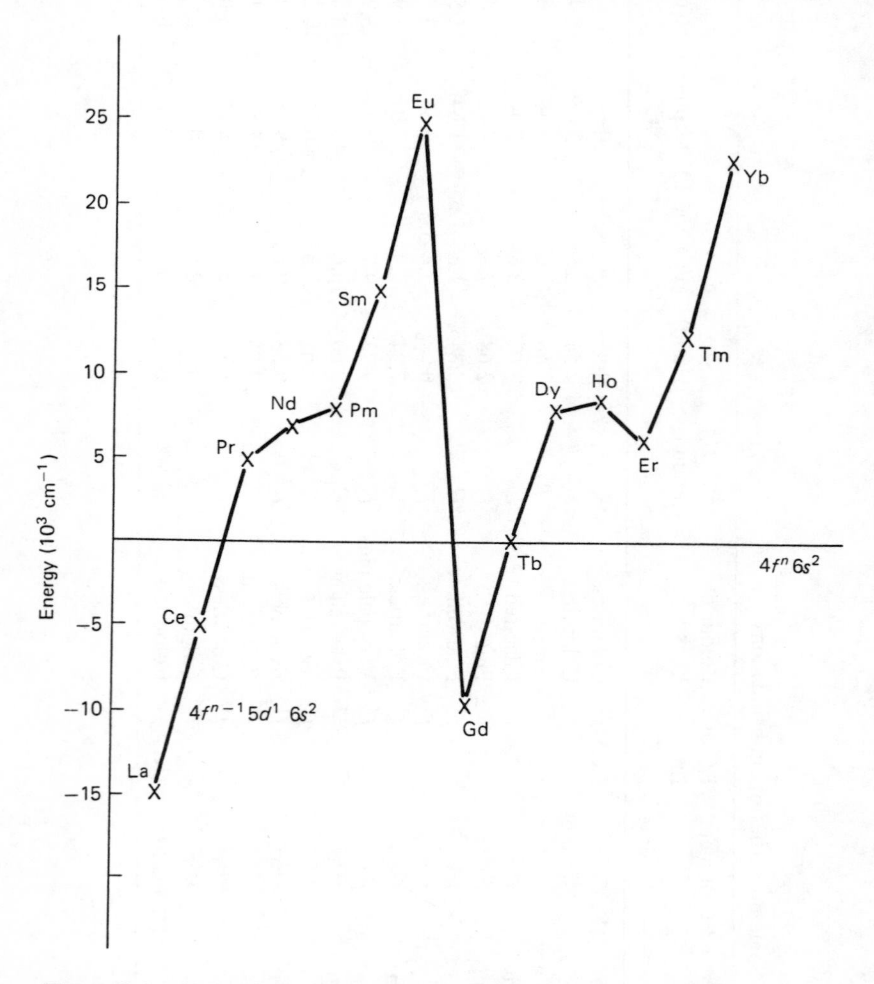

Figure 2.1 Approximate relative energies of the two possible ground state configurations of lanthanide atoms [based on M. Fred, *Advan. Chem. Soc.*, **71** (1967) 180; L. Brewer, in S. P. Sinha (ed.), *Systematics and Properties of the Lanthanides*, Reidel, Dordrecht, 1983, 17.69]

[Xe]$4f^7 6s^2$. At gadolinium however, the stability due to the set of half-filled $4f$ orbitals is such that the next electron is added to a $5d$ orbital, giving Gd the configuration [Xe]$4f^7 5d^1 6s^2$. This effect is short-lived, as the previous pattern is resumed from terbium, [Xe]$4f^9 6s^2$, until the $4f$ subshell is filled at ytterbium, so that lutetium, the last of this group of metals, has the configuration [Xe]$4f^{14} 5d^1 6s^2$. Yttrium has a structure analogous to lanthanum, [Kr]$4d^1 5s^2$.

As electrons are removed from the atoms to form the Ln^{3+} ions, they are taken preferentially from the $6s$ and $5d$ orbitals; at cerium, the stabilisation of $4f$ relative to $5d$ is now some 40 000–50 000 cm^{-1}, to judge from the transition in the UV region of the spectrum of the Ce^{3+} ion.

Conventionally, scandium, yttrium and lanthanum are placed together in Group 3 (IIIA in IUPAC form), with the lanthanide series following on lanthanum. The effect of the 'lanthanide contraction' means that yttrium is closer in size and properties to lutetium, so that a cogent argument has been advanced that places lutetium below yttrium, *preceded* by the series La–Yb. Thus lanthanum ([Xe]$6s^2 5d^1$) is followed by the lanthanides ([Xe]$4f^n 6s^2$) where the f subshell is filled, with lutetium ([Xe]$4f^{14} 6s^2 5d^1$) as the first member of the $5d$ block of metals. There are certainly similarities in the structures of many compounds of scandium, yttrium and lutetium; chlorides, cyclopentadienyls, and oxides, for example.

Chemically, yttrium has a strong resemblance to later lanthanides (such as dysprosium and erbium) as would be expected from its ionic radius.

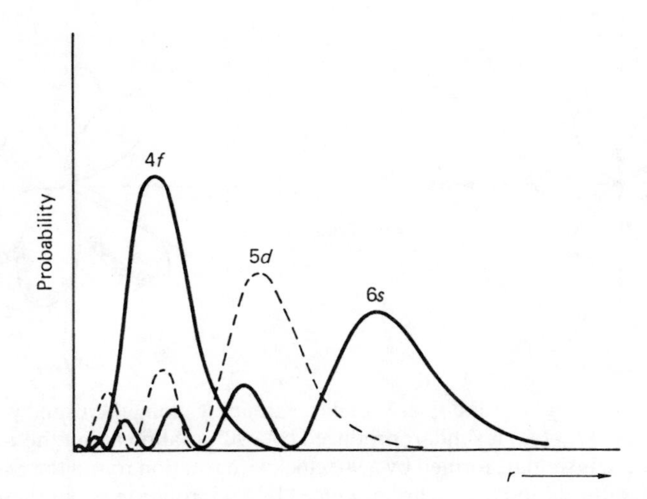

Figure 2.2 The radial part of the hydrogenic wave functions for the $4f$, $5d$ and $6s$ orbitals of cerium [after H. G. Friedman *et al.*, *J. Chem. Educ.*, **41** (1964) 357]

2.1.1 *f* orbitals

Figure 2.3 shows the cubic set of orbitals (showing the angular part of the hydrogenic wave functions). An alternative set (the 'general' set) is formed by a slightly different combination of four of the above orbitals.

The involvement of *f* orbitals in bonding in lanthanide and actinide compounds (and others) has been a controversial issue. It seems that, for the lanthanides, the regions of greatest 4*f* electron density simply do not extend far enough out from the nucleus (note Figure 2.2) for these electrons to take part in bonding to any great extent (the 5*f* electrons in the actinides may well be another matter!).

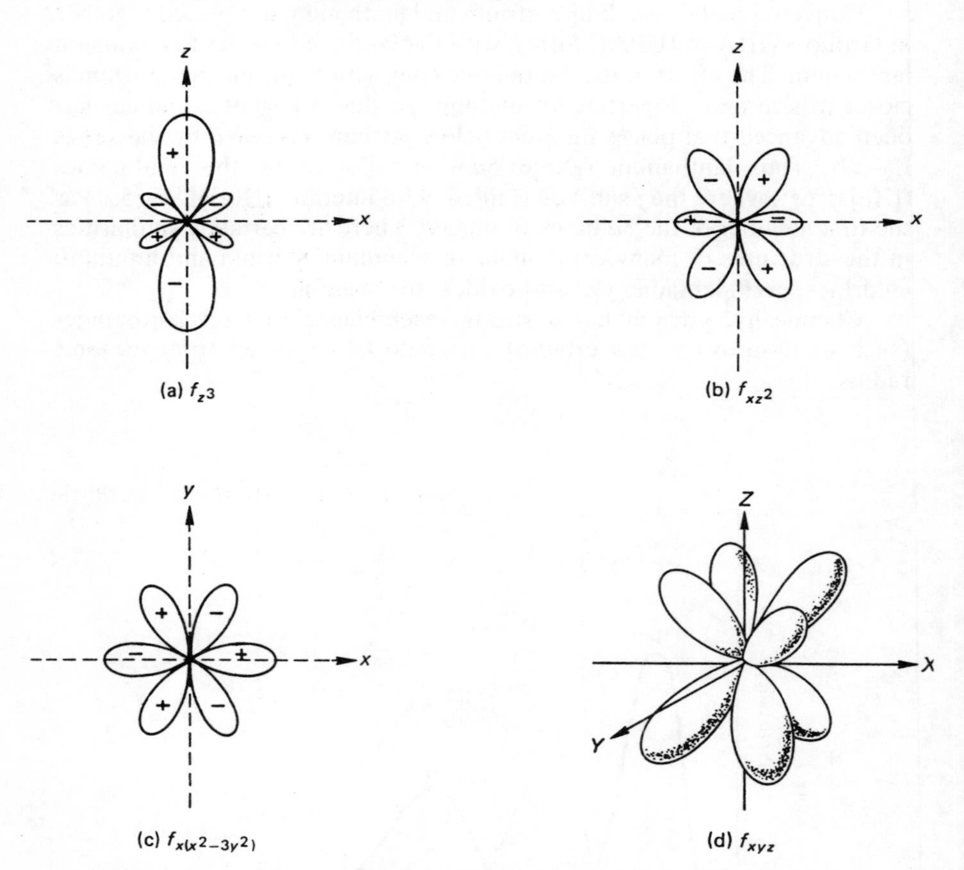

Figure 2.3 (a) f_{z^3} (f_{x^3} and f_{y^3} are similar, extending along the *x*- and *y*-axes respectively); (b) f_{xz^2} (f_{yz^2} is similar, produced by a 90° rotation about the *z*-axis); (c) $f_{x(x^2-3y^2)}$ $f_{y(3x^2-y^2)}$ is similar, formed by a 90° clockwise rotation round the *z*-axis); (d) f_{xyz} ($f_{x(z^2-y^2)}$, $f_{y(z^2-x^2)}$ and $f_{z(x^2-y^2)}$ are produced by a 45° rotation about the *x*, *y* and *z*-axes respectively). The cubic set comprises f_{x^3}, f_{y^3}, f_{z^3}, f_{xyz}, $f_{x(z^2-y^2)}$ $f_{y(z^2-x^2)}$ and $f_{z(x^2-y^2)}$; the general set is made of f_{z^3}, f_{xz^2}, f_{yz^2}, f_{xyz}, $f_{z(x^2-y^2)}$, $f_{x(x^2-3y^2)}$ and $f_{y(3x^2-y^2)}$

2.2 Ores: separation and processing

Table 2.2 lists some of the principal ores: these are generally found worldwide though most of the commercial extraction is carried out in the USA, especially from Californian bastnasite.

Table 2.2 Some rare earth ores*

Monazite	$CePO_4$
Xenotime	YPO_4
Gadolinite	$FeBe_2Y_2Si_2O_{10}$
Allanite	$(Ce,Ca)_2FeAlO(Si_2O_7)(SiO_4)(OH)$
Cerite	$Ce_3CaSi_3O_{13}H_3$
Euxenite	$Y_3(Nb,Ta)_3Ti_2O_{15}$
Bastnasite	$CeFCO_3$
Polycrase	$(Ce,Y,Th,U)(T,Nb,Ta)_2O_6$
Fluorocerite	CeF_3

*Ce represents a mixture of the lower lanthanides, especially La, Ce, Nd; Y represents a mixture of later lanthanides.

The abundance of individual lanthanides varies from one mineral to another; some, like monazite, are richer in the lighter elements, others (such as xenotime) contain more of the heavy ones, while a third class (such as euxenite) has a fairly even distribution. In all ores, however, the atoms with an even atomic number are more abundant (Figure 2.4) reflecting an enhanced nuclear stability (the Oddo and Harkins rule).

The traditional name 'rare earth' for these elements (which really refers to the oxides) is misleading, as even the rarest, lutetium, is more abundant than tantalum; neodymium is more abundant than gold or mercury.

Monazite and bastnasite are the principal ores worked. Before chemical treatment, the minerals are concentrated by processes like flotation to increase the lanthanide oxide content from about 7–10 per cent to 60 per cent. Monazite needs pre-treatment because of the thorium and phosphate content; traditionally this was achieved through digestion with H_2SO_4, turning the phosphates into soluble sulphates. Nowadays digestion with caustic soda is employed instead as it affords a better separation of the lanthanides and recovery of the phosphates, as well as eliminating corrosion problems and waste. To do this, the concentrated ore is heated with 65 per cent NaOH at 140° for some hours. On cooling, sodium phosphate is dissolved out with hot water and the hydroxides of thorium and the lanthanides are filtered off. The lanthanides are selectively dissolved out of this mixture with dilute HCl (or dilute HNO_3).

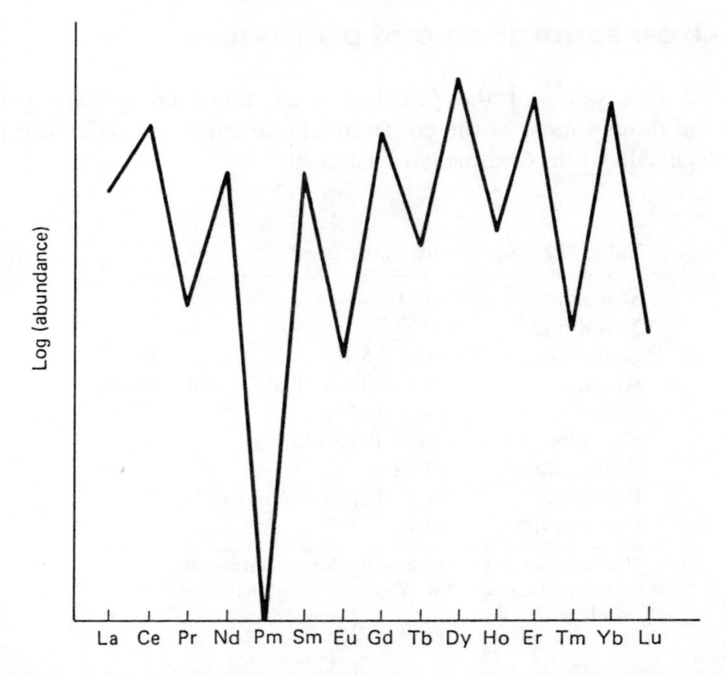

Figure 2.4 Distribution of lanthanides in euxenite, showing the increased abundance of nuclei with even atomic numbers

Acid leaching (dilute HCl) of bastnasite concentrates, to remove associated calcite, yields 70 per cent oxides. Some of this concentrate is mixed with carbon, then chlorinated forming mixed 'rare earth' chlorides – on electrolysis this forms the mixture of lanthanides called mischmetal. Otherwise the concentrate is treated with NaOH to turn it into a mixture of lanthanide hydroxides, which on dissolution in hydrochloric acid affords mixed lanthanide chlorides.

Separating the lanthanides

From the time (1794) that Gadolin obtained 'yttria', a mixture of lanthanide oxides, from Gadolinite, chemists were faced with the problem of separating 15 very similar elements. Selective reduction (Eu^{3+} to Eu^{2+}) or oxidation (Ce^{3+} to Ce^{4+}) could be exploited – thus $EuSO_4$ is insoluble unlike Ln^{3+} sulphates, for example – but until around 1950 most separations revolved around hundreds of tedious fractional crystallisations of compounds like the bromates, double sulphates ($NaLn(SO_4)_2.xH_2O$) or double nitrates (such as $Mg_3Ln_2(NO_3)_{12}.24H_2O$).

Ion-exchange methods, a spin-off from the Manhattan project, greatly simplified the separation; the mixed lanthanides were loaded on to a

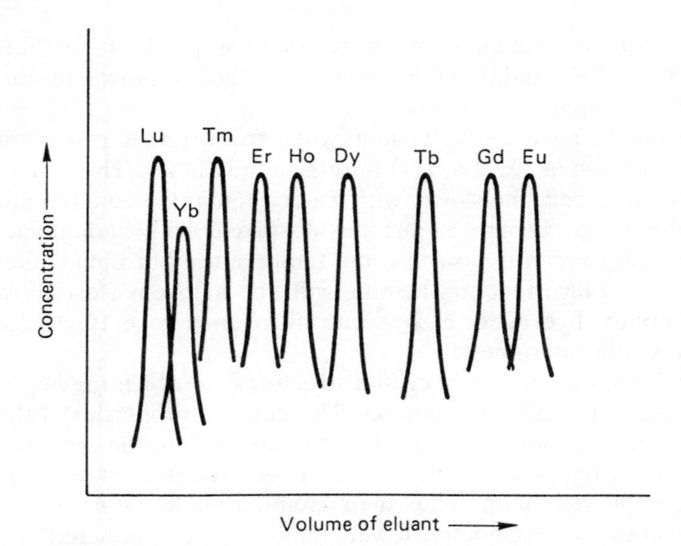

Figure 2.5 Separation of Ln^{3+} ions from an ion-exchange resin with a complexing agent like EDTA

cation-exchange resin then eluted with a suitable complexing agent like EDTA. The heavier, smaller lanthanide ions form stronger EDTA complexes so that they are removed from the resin first, thus the lanthanides are eluted in order of decreasing atomic number; with long columns, complete separations can be achieved (Figure 2.5).

Since the 1960s, solvent extraction has been the favoured process, employing complexing agents like tributylphosphate or di(2-ethylhexyl)phosphoric acid in a covalent solvent (such as kerosene).

2.3 The metals and their alloys

Individual rare earths are prepared by metallothermic (calcium) reduction of the anhydrous lanthanide fluoride or chloride:

$$LnF_3 \xrightarrow[1450^\circ]{Ca} Ln$$

The reaction is carried out under an argon atmosphere (at high temperature the lanthanides react with nitrogen) and the product is a calcium–lanthanide alloy, from which the calcium is removed by distillation. The metals with a tendency to the divalent state cannot be made this way (they are only reduced to the +2 state), but are obtained by lanthanum reduction of their oxides:

$$2La + M_2O_3 \rightleftharpoons La_2O_3 + 2M \qquad (M = Eu, Sm, Yb)$$

The divalent lanthanides are more volatile (bps of La, Eu, Sm, Yb are 3457, 1597, 1791, and 1193° respectively); their removal by distillation drives the reaction to completion.

The metals have a silvery shine when fresh, rather resembling steel, but several (such as Ce, La, Eu) tarnish rapidly in air. They are distinctly electropositive, reacting slowly with water even in the cold (compare Mg) and oxidising rapidly in moist air; they dissolve quickly in dilute acids. The reaction with oxygen is slow at room temperature but they ignite around 150–200°; they also react on heating with the halogens (fast above about 200°), sulphur, hydrogen, carbon and nitrogen (above 1000°) as well as with many other non-metals.

They adopt a number of crystal structures, of which the hcp structure of magnesium is the most common. The physical properties of the metals generally show a smooth trend across the series, with the exception of the metals with a tendency to form a stable +2 oxidation state, particularly europium and ytterbium. Thus their atomic radii are some 0.2 Å higher (and densities correspondingly lower) than would be predicted by extrapolation; likewise their melting points and boiling points are lower than expected (Figure 2.6). This can be rationalised in terms of a structure $Ln^{2+}(e^-)_2$ for europium and ytterbium, rather than $Ln^{3+}(e^-)_3$ adopted by other lanthanides, with the result that the outer electrons are not pulled in so near to the nucleus. The correspondingly weaker metallic bonding forces account for the higher volatility. The tendency to divalency in Eu and Yb is expressed in other ways, for example their solubility in liquid ammonia to form blue solutions of Ln^{2+}, and their reaction with organic halides to form Grignard-like species RMX.

The most familiar 'rare earth' alloy is mischmetal, a mixture of the lighter lanthanides (mainly cerium), made from unseparated lanthanide oxide mixtures. These are converted into anhydrous chloride, and electrolysed (graphite anode, iron cathode (container)) as a melt above about 820°. Mischmetal is used to make the ferrocerium alloy (75 per cent mischmetal, 25 per cent iron) employed for lighter flints and which is also used in iron foundries for deoxidation and desulphurisation of steels. Alloying with small amounts of lanthanides often improves mechanical properties and corrosion resistance.

The alloy $LaNi_5$ has attracted considerable attention for its ability to absorb H_2 (to afford $LaNi_5H_6$); hydrogen is desorbed fast at 140° and thus the alloy has potential for H_2 storage and removal of H_2 from gas mixtures. Other applications involve hydrogen storage electrodes and fuel cells, as well as a hydrogenation catalyst for the reduction of alkenes, alkynes, aldehydes, ketones and nitriles. $CeNi_5$ has been studied as an active catalyst for the hydrogenation of CO. M_2Co_7 (M = Nd, Sm) dehydrogenate organic compounds under relatively mild conditions. $SmCo_5$ has been widely used as an outstanding material for making permanent magnets.

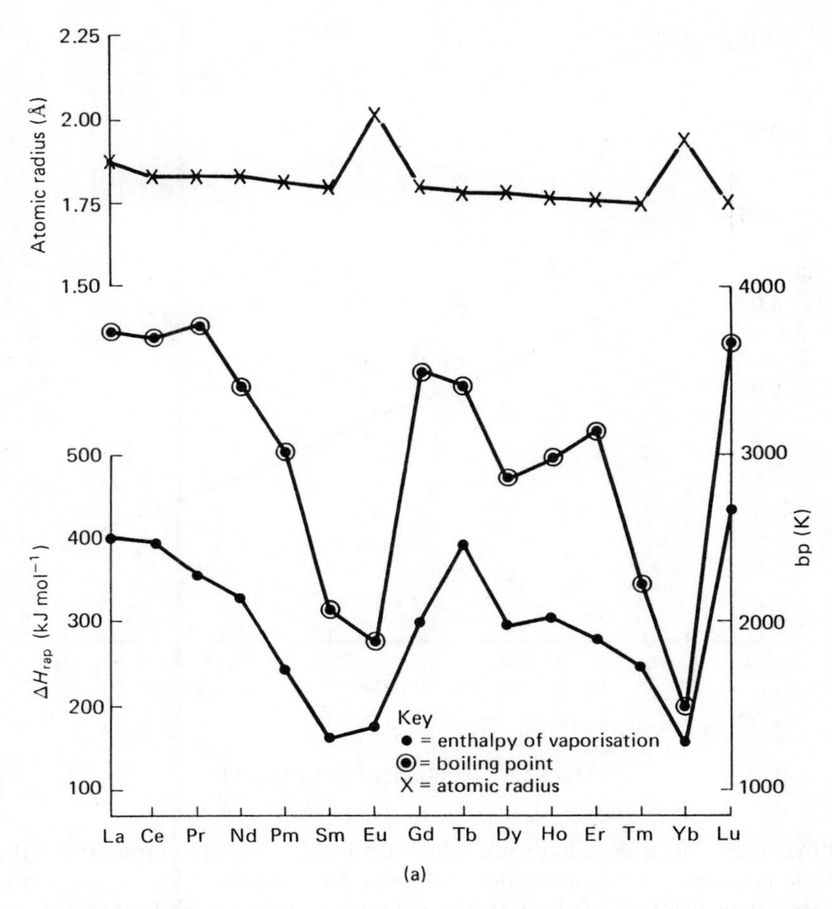

Figure 2.6 (a) Trends in some properties of lanthanide atoms. (b) Ionic radii of lanthanide (Ln^{3+}) and actinide (An^{3+}) ions (data for 6-coordination)[see page 20]. Note the regularity of change for Ln^{3+} ions, in contrast to the metals

Hydrides

As mentioned above, alloy hydrides like $LaNi_5H_6$ and $AlLaNi_4H_4$ have attracted attention for hydrogen storage and fuel cell electrodes. The pure metals, however, readily form hydrides, by direct reaction of the elements, usually at around 300°. There are two main series, MH_2 (fluorite structure) and MH_3, though they are frequently non-stoichiometric, for example, lutetium hydride has existence ranges $LuH_{1.83-2.23}$ and $LuH_{2.78-3.00}$. Europium, as might be expected for a metal with a stable +2 state, only forms EuH_2. Low hydrogen pressures are needed to ensure formation of the

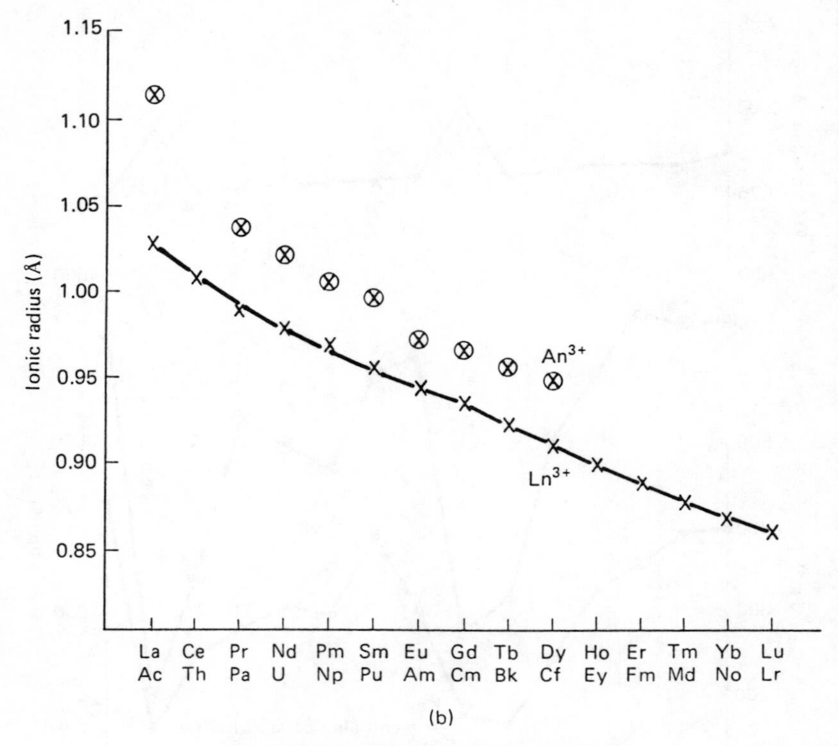

Figure 2.6 (continued) [see page 19]

dihydrides, but unless high pressures are used, a substoichiometric trihydride is formed. The dihydrides are generally good conductors, suggesting a structure $M^{3+}(H^-)_2(e^-)$ with the trihydrides being salt-like $M^{3+}(H^-)_3$.

2.4 Characteristics of the lanthanides

These include:

1 A similarity in physical properties throughout the series, with changes tending to occur as gradual trends.

2 The adoption of predominantly the +3 oxidation state for all the elements.

3 The adoption of coordination numbers greater than 6 (usually 8–9) in their compounds.

4 Coordination polyhedra largely determined by steric rather than crystal-field considerations; non-directional bonding.

5 A tendency to decreasing coordination number across the series as the radius of the lanthanide ion decreases.

6 A preference for 'hard' donor atoms (such as O, F), binding in order of electronegativity.

7 Very small crystal-field effects.

8 Sharp 'line-like' electronic spectra of complexes which show little dependence on ligands.

9 Magnetic properties of a given ion again largely independent of environment.

10 Ionic complexes undergo rapid ligand-exchange.

2.5 Stability of lanthanide oxidation states

It is a fact that in aqueous solution, the predominant oxidation state of the lanthanides is +3 – cerium(IV) ions are metastable; europium(II) ions are fairly stable in aqueous solution. Even in solid compounds, their chemistry is largely that of the tripositive state.

The stability of compounds in a particular oxidation state depends on the interrelationship between a number of factors; to talk simply in terms of 'half-filled' shells or electron configuration is facile.

2.5.1 Lanthanide halides

Why does lanthanum only form the trifluoride LaF_3, and not LaF_2 or LaF_4? Thermodynamic data on the lanthanides are scarce; to simplify the matter, entropy changes will be neglected, so that free energy changes are replaced by enthalpy changes.

To estimate the enthalpies of formation of the fluorides requires the enthalpy of atomisation and ionisation enthalpies of lanthanum, the dissociation enthalpy and electron affinity of fluorine, and appropriate lattice enthalpies. An experimental lattice enthalpy is available for LaF_3; there are calculated values available for LaF_2 and LaF_4, alternatively approximate values can be obtained by using those for BaF_2 (-2350 kJ mol^{-1}) and CeF_4 (-8391 kJ mol^{-1}) respectively. Thus for LaF_4:

	ΔH (kJ mol^{-1})
$La(s) \longrightarrow La(g)$	423
$La(g) \longrightarrow La^{4+}(g) + 4e$	538 + 1067 + 1850 + 4820
$2F_2(g) \longrightarrow 4F(g)$	336
$4F(g) + 4e^- \longrightarrow 4F^-(g)$	−1312
$La^{4+}(g) + 4F^-(g) \longrightarrow LaF_4(s)$	−8391
$La(s) + 2F_2(g) \longrightarrow LaF_4(s)$	−629

Similar calculations yield $\Delta H_f(LaF_3) = -1726$ kJ mol^{-1} and $\Delta H_f(LaF_2) = -812$ kJ mol^{-1}. Thus LaF_2 and LaF_4 are exothermic with respect to the elements, but in assessing their stability, other decomposition pathways need to be assessed. LaF_4 may decompose:

$$LaF_4 \longrightarrow LaF_3 + \tfrac{1}{2}F_2$$

and it can be seen that the more exothermic ΔH_f for LaF_3 implies that this pathway would be very feasible. As far as LaF_2 is concerned, a disproportion pathway should be considered:

$$3LaF_2(s) \longrightarrow 2LaF_3(s) + La(s)$$

ΔH can be assessed using Hess's Law at −1016 kJ mol^{-1}. This simplified model thus indicates that both LaF_4 and LaF_2 are unstable with respect to LaF_3; in the case of LaF_4, the high I_4 is possibly the most important single factor in its instability.

Fluorine, however, is the halogen most capable of promoting a high oxidation state, the high lattice energies associated with solid compounds of the small fluoride ion being important. In the case of cerium, using experimentally determined lattice enthalpies, a different picture emerges:

	U_L (= ΔH lattice)	$\Delta H_f(CeX_n)$
CeF_3	−4915	−1719
CeF_4	−8397	−1903

Here the lower I_4 of Ce (3547 kJ mol^{-1}) and the high lattice enthalpy of CeF_4 are factors promoting the stability of CeF_4, which may in fact be synthesised. ΔH for the process $CeF_4 \rightarrow CeF_3 + \tfrac{1}{2}F_2$ is now positive, indicating that ΔG may be positive too. A similar calculation performed for PrF_3 and PrF_4 yields $\Delta H_f(PrF_3) = -1697$ kJ mol^{-1} and $\Delta H_f(PrF_4) = -1674$ kJ mol^{-1} (other estimates are −1712 and −1690 kJ mol^{-1} respectively). This suggests that PrF_4 is unstable with respect to PrF_3 and F_2 – in fact, PrF_4 (which decomposes at around 70°) cannot be made by fluorination of PrF_3, but by the action of dry HF on Na_2PrF_6. $CeCl_4$, the most plausible tetrachloride, does not exist; although exothermic (ΔH_f = −820 kJ mol^{-1}), it is not as stable as $CeCl_3$ (ΔH_f = −1058 kJ mol^{-1}). Salts of the $CeCl_6^{2-}$ ion can be made, presumably the greater lattice energy contributing to their stability.

Similar calculations can be carried out for compounds in the dipositive state. Thus for europium, using experimental enthalpies of formation for EuF_2 and EuF_3 of −1188 and −1584 kJ mol^{-1} respectively, ΔH calculated for the process $3EuF_2 \rightarrow 2EuF_3 + Eu$ is +396 kJ. Europium is favoured in the divalent state particularly on account of its high I_3 and low enthalpy of atomisation. The divalent state will also be favoured by low lattice energies (for example, iodides).

Nearly all trihalides are known, the exception being EuI_3. Its instability with respect to EuI_2

$$EuI_3 \longrightarrow EuI_2 + \tfrac{1}{2}I_2$$

is supported by calculated (literature) values of ΔH_f of −540 kJ mol^{-1} and −630 kJ mol^{-1} for EuI_3 and EuI_2 respectively. Another decomposition pathway is the reproportionation

$$Ln + 2LnX_3 \longrightarrow 3LnX_2$$

Application of Hess's Law suggests that this process will be endothermic unless $\dfrac{\Delta H_f(LnX_3)}{\Delta H_f(LnX_2)} < 1.5$.

This reaction is used to synthesise those dihalides of marginal stability that cannot be made by hydrogen reduction of the trihalides, by heating a mixture of the molten trihalides and metal, followed by rapid quenching; for example

$$Nd + 2NdCl_3 \longrightarrow 3NdCl_2$$

In general, the existing dihalides correlate well with the value of I_3 for the metal: europium (and others with high third ionisation enthalpies) are more ready to form them.

2.5.2 The aquo-ions

The pronounced stability of Ln^{3+}(aq) suggests that both these processes are generally favourable:

$$Ln^{2+}(aq) + H^+(aq) \longrightarrow Ln^{3+}(aq) + \tfrac{1}{2}H_2(g)$$

$$2Ln^{4+}(aq) + H_2O(l) \longrightarrow 2Ln^{3+}(aq) + 2H^+(aq) + \tfrac{1}{2}O_2(g)$$

(Thus Ln^{2+} reduces water and Ln^{4+} will oxidise it.)

We can set up a thermodynamic cycle (again neglecting entropy contributions):

$$\begin{array}{ccc} Ln^{2+}(g) & \xrightarrow{\Delta H_{hydr}(Ln^{2+})} & Ln^{2+}(aq) + H^{+}(aq) \\ \downarrow I_3 & & \downarrow \Delta H_{ox}(Ln^{2+}) + \Delta H_H \\ Ln^{3+}(g) & \xrightarrow{\Delta H_{hydr}(Ln^{3+})} & Ln^{3+}(aq) + \tfrac{1}{2}H_2(g) \end{array}$$

$$\therefore \Delta H_{ox} = I_3 + (\Delta H_{hydr}Ln^{3+} - \Delta H_{hydr}Ln^{2+}) - 439 \text{ kJ mol}^{-1}$$

In others words, the stability of Ln^{2+}, favoured by a positive value of ΔH_{ox}, is more likely with a high third ionisation enthalpy and a small difference in the (exothermic) enthalpies of hydration.

Correspondingly, the stability of the +4 oxidation state will be determined by the relationship between the fourth ionisation enthalpy and the difference between the hydration enthalpies.

Hydration enthalpies increase across the series as the ion size decreases, and will thus be very similar for adjacent lanthanides. Some literature values (estimated for some dipositive ions) are −3278, −3449, −3501, −3519 and −3706 kJ mol^{-1} for La^{3+}, Sm^{3+}, Eu^{3+}, Gd^{3+} and Yb^{3+} respectively, and −1327, −1444, −1458, −1472 and −1594 kJ mol^{-1} for the corresponding dipositive ions. Substitution of these values in the above equation leads to ΔH_{ox} values of −540 kJ mol^{-1} for La^{2+} and −78 kJ mol^{-1} for Eu^{2+}, correlating with the increased (but marginal) stability of Eu^{2+} over other lanthanide(II) ions.

2.5.3 Patterns in ionisation energies

Figure 2.7 represents variations in ionisation energies across the series, representing cumulative ionisation energies. Some correlation is evident, with the stability of the half-filled and filled shells explicable in terms of exchange energy, which depends on the number of electrons with parallel spins. Note the low I_3 for Gd and high I_3 for Eu, for example; as we have seen there is a correlation between the magnitude of I_3 and the stability of Ln^{2+} in compounds like the dihalides. Furthermore, the irregularities in the pattern are reinforced by the fact that the third electron removed from both Gd and Lu does not come from a 4*f* orbital, as it does for the other lanthanides. Apart from the discontinuity at gadolinium, there are other discontinuities in I_3 (and other quantities dependent on I_3) at the quarter – and three-quarter – shells, that have been attributed to variation in interelectronic repulsive forces. Ionisation energies are listed in Appendix A.

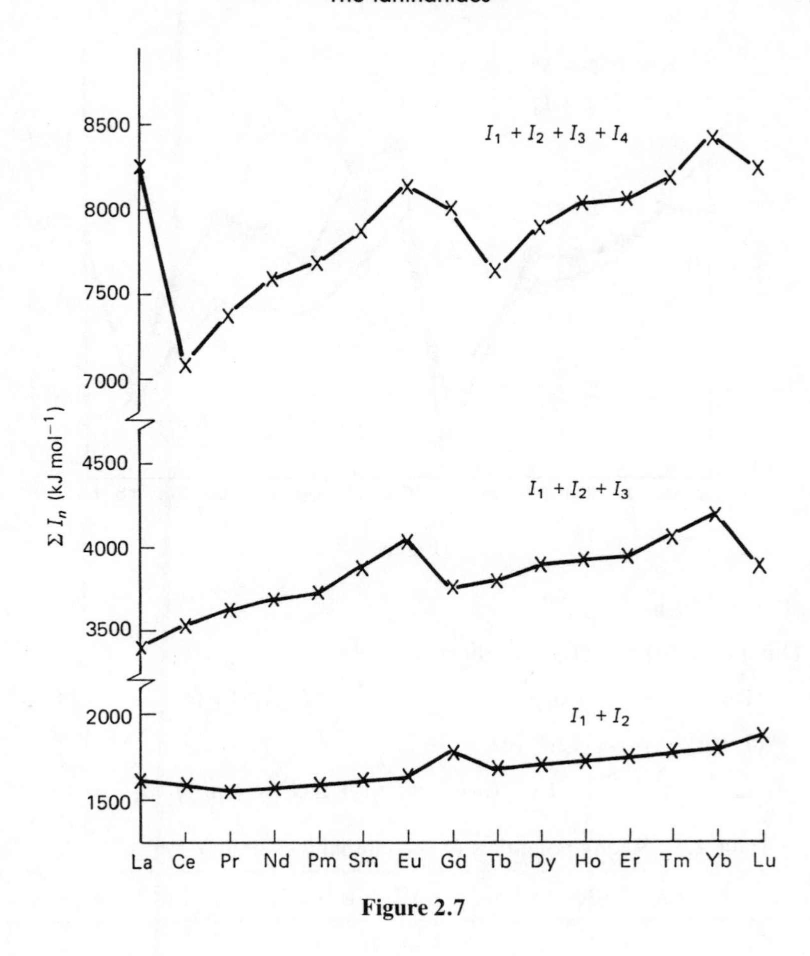

Figure 2.7

2.5.4 Patterns in electrode potentials

These are collected in Table 2.3: many of the values involving the less stable oxidation states are estimates.

(i) Ln^{3+}/Ln^{2+}

Since I_3 relates directly to ΔH_{ox} for the process $Ln^{2+} \rightarrow Ln^{3+}$, and $\Delta G = -nFE$, it is therefore not surprising that the trend in E^0 mirrors the pattern in I_3 (Figure 2.8). As I_3 increases from the beginning of the series, the stability of Ln^{2+} reaches a peak at Eu^{2+}; the sharp drop in I_3 on passing to gadolinium, the next element, means that the dipositive state disappears. The trend is repeated in the second half of the series, with the moderately stable Yb^{2+} being followed by the non-existent Lu^{2+}.

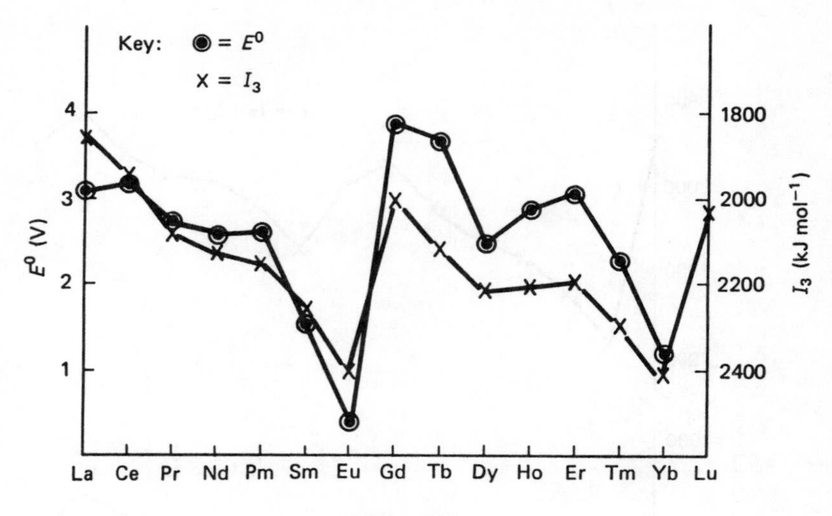

Figure 2.8

(ii) Ln^{3+}/Ln

This potential depends on these processes:

$$Ln(s) \longrightarrow Ln(g) \qquad \Delta H_{atom}(Ln)$$

$$Ln(g) \longrightarrow Ln^{3+}(g) + 3e^- \qquad I_1 + I_2 + I_3$$

$$Ln^{3+}(g) \longrightarrow Ln^{3+}(aq) \qquad \Delta H_{hydr}(Ln^{3+})$$

Table 2.3 Redox potentials of the lanthanides (volts)

	$Ln^{3+} + 3e^- \rightarrow Ln$	$Ln^{3+} + e^- \rightarrow Ln^{2+}$	$Ln^{4+} + e^- \rightarrow Ln^{3+}$
La	−2.37	(−3.1)	
Ce	−2.34	(−3.2)	1.7
Pr	−2.35	(−2.7)	(3.4)
Nd	−2.32	−2.6 (in THF)	(4.6)
Pm	−2.29	(−2.6)	(4.9)
Sm	−2.30	−1.55	(5.2)
Eu	−1.99	−0.34	(6.4)
Gd	−2.29	(−3.9)	(7.9)
Tb	−2.30	(−3.7)	(3.3)
Dy	−2.29	−2.5 (in THF)	(5.0)
Ho	−2.33	(−2.9)	(6.2)
Er	−2.31	(−3.1)	(6.1)
Tm	−2.31	−2.3 (in THF)	(6.1)
Yb	−2.22	−1.05	(7.1)
Lu	−2.30		(8.5)
Y	−2.37		

Values in parentheses are calculated values.

The first two of these are endothermic and the third is exothermic, so that the overall enthalpy change is the difference between two large numbers. Despite the inherent uncertainty, the overall ΔH remains remarkably constant across the series, typical values being 608 (La), 630 (Gd), 613 (Tm) and 593 kJ mol^{-1} (Lu), apart from ytterbium (644 kJ mol^{-1}) and europium (712 kJ mol^{-1}), caused by the high I_3 values for these 'divalent' metals. The effect of this on E^0 can readily be seen from Table 2.3.

2.5.5 Variations in energy change across the series

If ΔH (or ΔG) values for a particular process are plotted across the series, three patterns of behaviour are observed.

In cases where there is no change in oxidation state (*f* electrons are 'conserved'), a smooth variation is seen for complex-forming reactions like

$$Ln^{3+}(aq) + EDTA^{4-}(aq) \longrightarrow Ln(EDTA)^{-}(aq)$$

or

$$Ln^{3+}(g) \longrightarrow Ln^{3+}(aq)$$

A second class involves those where there are slight discontinuities at europium and ytterbium, as in the formation of the trichlorides, because the electronic structures of the elements Eu and Yb are slightly different from those of the other lanthanides. Finally, there are processes where the variation resembles the variation in I_3, for processes like

$$2LnCl_2 + Cl_2 \longrightarrow 2LnCl_3$$

Here the *f* electron population is decreasing by one, and the relationship with I_3 is plausible.

2.6 Spectroscopic and magnetic properties of the lanthanides

2.6.1 Introduction

This section is concerned largely with properties of the Ln^{3+} ions, which are described accurately by the Russell–Saunders coupling scheme. In this, the spins of individual *f*-electrons (s) are coupled together to afford the spin quantum number for the ion (S). Similarly, the orbital angular momenta, l, of the individual electrons are added vectorially. For an *f* electron, $l = 3$, so that m_l can have any of the seven integral values between $+3$ and -3. Vectorial addition of the m_l values for a multielectron ion affords L, the total orbital angular momentum quantum number:

L	0	1	2	3	4	5	6	7
State symbol	S	P	D	F	G	H	I	K

A weaker coupling, spin orbit coupling, exists between S and L. The values of the resulting quantum number, J, are obtained by vector addition of L and S. Thus for the I state of Pm^{3+}: $2S + 1 = 5$, therefore $S = 2$, and $L = 6$ (I state). Hence J can have values of $(L + S),\ldots,(L - S)$, that is 8, 7, 6, 5 and 4.

The ground state for the ion is determined by Hund's rules. For an ion $^{2S+1}L_J$:

1 The spin multiplicity $(2S + 1)$ is the highest possible.

2 If there is more than one term with the same spin multiplicity, the term with the highest L value is the ground state.

3 For a shell less than half-filled, J is as low as possible; for a shell more than half-filled, J is as high as possible.

Taking an f^3 ion as an example, the ground state will have $S = \frac{3}{2}$, therefore $(2S + 1) = 4$; $L = 3 + 2 + 1 = 6$, hence it will be an I state ion. J can have values of $(6 + \frac{3}{2})$, $(6 + \frac{3}{2}) - 1,\ldots,(6 - \frac{3}{2})$, that is $\frac{15}{2}$, $\frac{13}{2}$, $\frac{11}{2}$ or $\frac{9}{2}$. Since the ion has less than a half-filled shell, J will be $\frac{9}{2}$ for the ground state, so that the ground state of Nd^{3+} is $^4I_{9/2}$.

The ground states for all Ln^{3+} ions are listed in Table 2.1. Theoretical predictions (Hartree–Fock calculations including configuration interaction) and comparison with experiment have resulted in term schemes for the lanthanides of the type shown in Figure 2.9.

The energy levels of the rare earth ions in halide lattices (such as LnF_3), determined accurately by spectroscopy, strongly resemble those of the gaseous ions. The reason for this is that the filled 'outer' 5*s* and 5*p* orbitals largely shield the 4*f* electrons from surrounding ligands, so that crystal field splittings amount at most to some 100 cm^{-1}, and thus the weak CF splittings may be regarded as a perturbation on the free-ion energy levels. (In contrast, in the 3*d* transition metals, crystal-field splittings are large and *L–S* coupling weak). The lack of crystal field effects contributes to the *f–f* absorption bands in compounds being nearly as narrow as in gaseous ions.

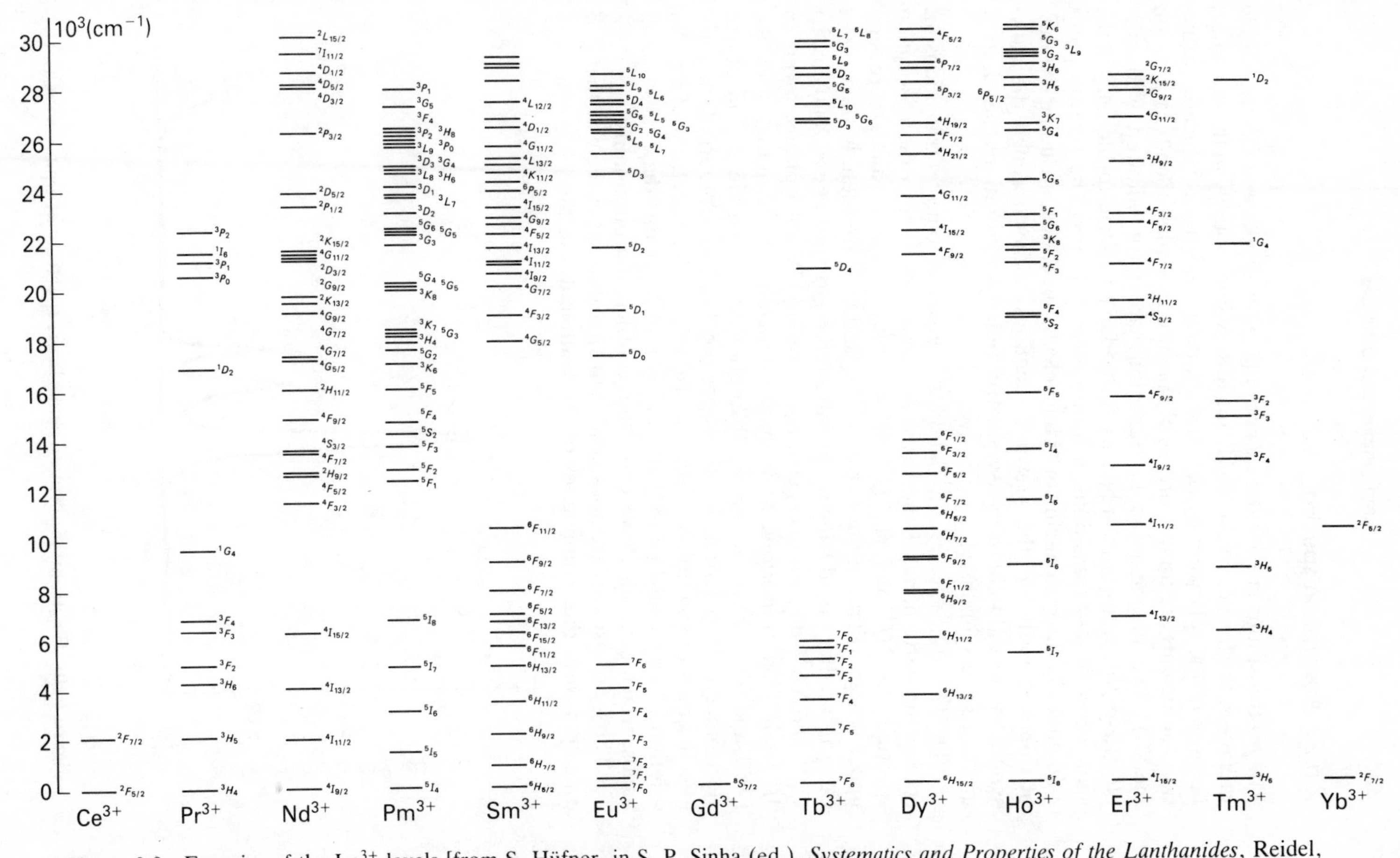

Figure 2.9 Energies of the Ln^{3+} levels [from S. Hüfner, in S. P. Sinha (ed.), *Systematics and Properties of the Lanthanides*, Reidel, Dordrecht, 1983; reproduced by permission]

2.6.2 Absorption spectra

The *f–f* transitions are excited by both magnetic-dipole and electric-dipole radiation, the former transitions being observable, notably in fluorescence, because of the relatively weak electric-dipole transitions (typically extinction coefficients are up to 2 orders of magnitude weaker than for transition metals). The magnetic-dipole transitions are parity-allowed while the electric-dipole transitions are parity-forbidden ('Laporte-forbidden') in the same sense as *d–d* transitions in transition-metal ions. The *f–f* transitions gain intensity through mixing in of high electronic states, including *d* states, of opposite parity to the 4*f* wave functions, either through the (low-symmetry) ligand field or by asymmetric molecular vibrations that destroy any centre of symmetry.

The f^0 and f^{14} species, La^{3+} and Lu^{3+}, give rise to no *f–f* transitions, of course; similarly none of the sharp *f–f* bands are seen in the absorption spectra of $Ce^{3+}(f^1)$ and $Yb^{3+}(f^{13})$ as, with only a single *L* value, there is no upper 4*f* state. (Transitions between $^2F_{5/2}$ and $^2F_{7/2}$ are seen, in the case of Ce^{3+} as a rather broad band in the infra-red region just above 2000 cm^{-1}.) Ce^{3+} and Yb^{3+} do, however, give rise to broad UV absorptions, owing to $4f^n \rightarrow 4f^{n-1}5d^1$ transitions, as do many lanthanides. Certain Ln^{3+} ions like Eu^{3+} and Tb^{3+} have no strong absorption in the visible region of the electromagnetic spectrum, and thus appear colourless. Colours of the Ln^{3+} ions in aqueous solution are listed in Table 2.1. The spectrum of Pr^{3+} in solution is shown in Figure 2.10, and is typical of the sharp absorption bands, which generally have extinction coefficient around unity.

The *f–f* transitions are not sensitive to environment in the way that *d–d* transitions are, thus compounds of the lanthanides in the (+3) oxidation

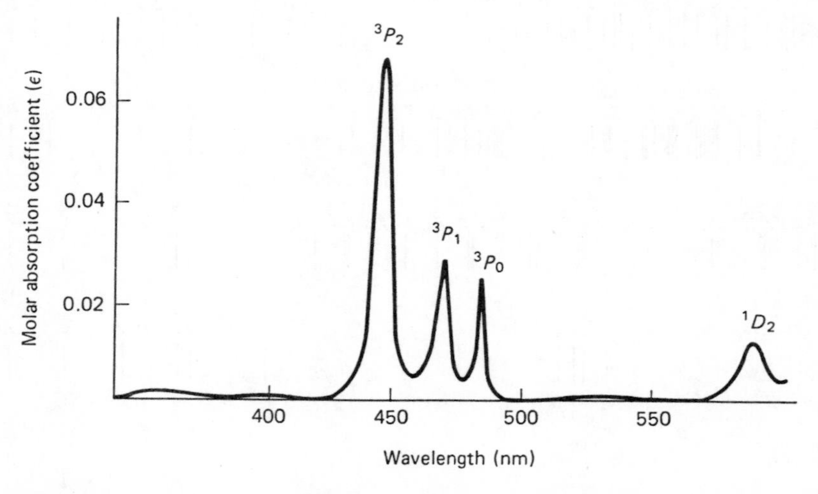

Figure 2.10 Electronic absorption peaks in the spectrum of aqueous $PrCl_3$

state are usually much the same colour as the aquo-ions. Certain absorption bands, particularly in compounds of Nd^{3+}, Ho^{3+} and Er^{3+}, display 'hypersensitivity' of position and profile to environment, discussed in section 2.7.1.

Most Ln^{2+} ions are unstable in solution (particularly in water) but have been stabilised in CaF_2 lattices, prepared by doping the lattice with Ln^{3+}, then γ-irradiating. The difference between corresponding levels is less than in the isoelectronic Ln^{3+} ions because of the reduced charge, and thus the ions have different colours (for example, Eu^{2+} yellow *versus* Gd^{3+} colourless).

2.6.3 Lanthanide luminescence (fluorescence) spectra

Ultra-violet irradiation of many lanthanide complexes, particularly with conjugated organic ligands, causes fluorescence, the emitted light exhibiting sharp lines characteristic of the $4f$–$4f$ transitions of the Ln^{3+} ion.

The generally accepted mechanism is as follows (Figure 2.11). An electron is promoted to an excited ligand singlet state. It may either return direct to the ground state (ligand fluorescence) or follow a non-radiative pathway to a ligand triplet state. From the triplet state, the electron can drop directly back to the ground state (phosphorescence) or again undergo non-radiative intersystem crossing, this time to a nearby excited state of a Ln^{3+} ion. From here it returns to the ground state either by non-radiative emission or by metal–ion fluorescence. Certain lanthanide ions (Sm^{3+}, Dy^{3+}, and particularly Tb^{3+} and Eu^{3+}) have excited states lying slightly lower in energy than the triplet states of typical ligands, hence exhibiting strong metal–ion fluorescence. Of the other Ln^{3+} ions, La^{3+} and Lu^{3+} have no f^n excited state. $Gd^{3+}(f^7)$ has all its excited states *above* the ligand triplet states, while the other ions have a large number of excited states, promoting energy loss by a non-radiative route.

Tb^{3+} and Eu^{3+} are the two most useful ions for these studies. They fluoresce with green and red colours respectively: for Tb^{3+}, the main emissions are ${}^5D_4 \rightarrow {}^7F_n$ (n = 6–0) with the ${}^5D_4 \rightarrow {}^7F_5$ the strongest, while for Eu^{3+} ${}^5D_0 \rightarrow {}^7F_n$ (n = 4–0) are seen, with the transitions to ${}^7F_0$, ${}^7F_1$ and ${}^7F_2$ the most useful. Owing to the weak nature of the electric-dipole transitions, the magnetic-dipole transitions are of comparable intensity. (For Eu^{3+}, ${}^5D_0 \rightarrow {}^7F_x$ are predominantly electric dipole for x even, magnetic dipole for x odd.) The ligand field in a complex removes the J degeneracy of a given ${}^{2S+1}L_J$ term partly or completely, the extent of this (and the resultant splitting of the emission line) depending on the symmetry of the ligand field. The application of this to the study of biological systems and the determination of the symmetry of complexes is discussed in sections 2.7.2. and 2.7.9.

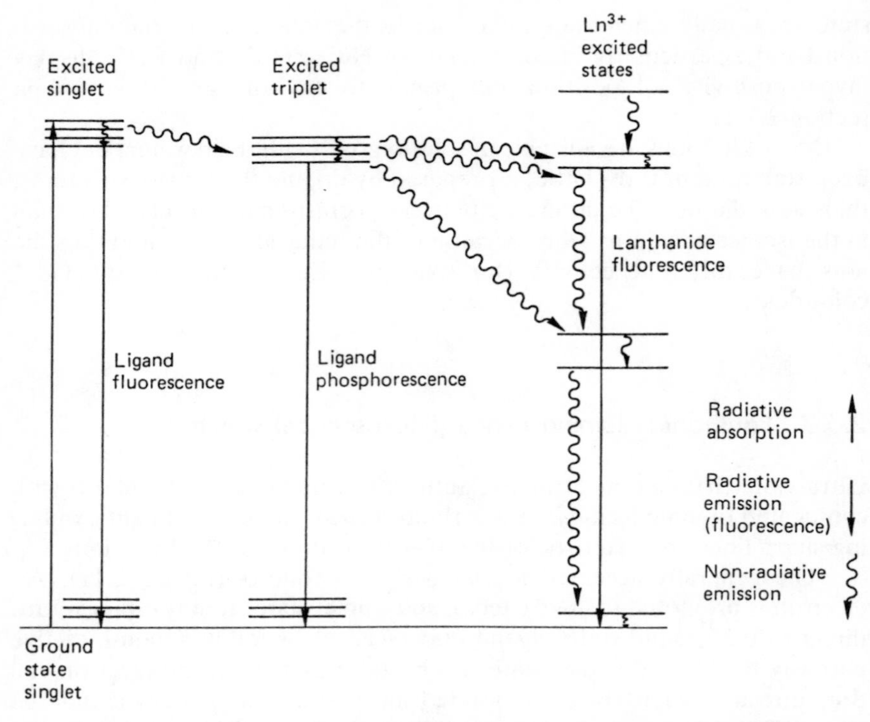

Figure 2.11 Quantitative energy level diagram for a lanthanide complex showing transitions from ligand and metal-ion excited states

2.6.4 Applications of rare earth ions to colour television

This is the largest market for rare earth phosphors. A traditional colour TV tube employs three electron guns to fire electrons at the screen from subtly different angles (through a 'mask' of some kind) to hit clusters of three phosphor 'dots', each 'dot' emitting a different primary colour. Early red phosphors were broad-band emitters like Mn^{2+}–activated phosphate and Ag:Zn,CdS, replaced in the mid-1960s by Eu^{3+}:YVO_4 and, subsequently, by Eu^{3+} in Y_2O_2S or Eu^{3+}:Y_2O_3 which, because of their narrow bands, are brighter, even though less energetically efficient, than the broad-band phosphors. They also match the eye's colour response better. For green, there is a choice between broad-band emitters Cu,Al:Zn,CdS or the 5*d*–4*f* emitters Ce^{3+}:CaS and Eu^{2+}:$SrGa_2S_4$, against the 'narrow line' 4*f*–4*f* emitter Tb^{3+}:La_2O_2S. The 'standard' blue is Ag,Al:ZnS and no rare-earth emitter such as Tm^{3+}:ZnS is capable of replacing it.

2.6.5 Neodymium lasers

Neodymium ions in a solid matrix can be made to exhibit laser action, the most popular being Nd^{3+} in yttrium aluminium garnet (YAG). (Laser is an acronym for *L*ight *a*mplification by *s*timulated *e*mission of *r*adiation. It relies on increasing the emission of light by stimulating the release of photons from, in this case, excited Nd^{3+} ions.)

In a typical device, a YAG rod a few centimetres long containing around 1 per cent Nd^{3+} is fitted with a mirror at each end, one being a partly transmitting mirror (or similar device). On irradiation with a suitable lamp (such as W/I_2) the system is 'pumped' to ensure that an excess of Nd^{3+} ions is in an excited state, such as $^4F_{5/2}$ and $^4F_{7/2}$ (so that more ions can emit than absorb); these decay rapidly (cascade) to the long-lifetime $^4F_{3/2}$ state, thus a large fraction of the Nd^{3+} ions are in this state compared with the ground state (this is called 'population inversion'). If a photon of the right energy (wavelength of the laser transition) hits a Nd^{3+} ion, the Nd^{3+} ion is stimulated to release a second (as it drops to the $^4I_{11/2}$ state). As the photons are reflected to and fro in the rod, more and more ions are stimulated into giving up photons and eventually the build-up of photons is so great that they emerge from the rod as an intense beam of monochromatic light, wavelength 1.06 μm (near infra-red).

The $^4I_{11/2}$ state is an excited level of the ground state which is not thermally populated, and undergoes rapid relaxation to the $^4I_{9/2}$ ground state. The neodymium thus acts as a 'four-level' laser (Figure 2.12).

2.6.6 Magnetic properties

These are determined by the 4*f* electrons, which do not interact with their ligands and are largely unaffected by their environment, hence their orbital

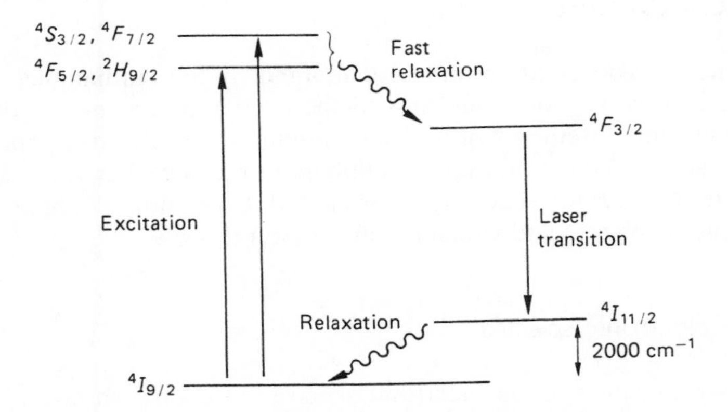

Figure 2.12 Laser action in Nd^{3+}: YAG (crystal-field splitting not shown)

moment is unquenched. The magnetic properties are thus effectively those of the free ions, varying little from one compound to another (Table 2.1).

The susceptibility is given by

$$\chi_{\mathrm{m}} = \frac{Ng^2\beta^2 J(J+1)}{3kT}$$

and the magnetic moment by

$$\mu_{\mathrm{eff}} = g_J\sqrt{[J(J+1)]}$$

where

$$g_J = \frac{S(S+1) + 3J(J+1) - L(L+1)}{2J(J+1)}$$

Application of the simple formula, due to Hund, leads to the moments listed in Table 2.1. The higher values for the second part of the lanthanide series should be noted, and are due to $J = L + S$ for f^n with $n > 7$(Hund's third rule). In practice there is good agreement with experimental values, except for compounds of Sm^{3+} and Eu^{3+}. Van Vleck showed that this deviation arose from neglecting low-lying states – $^7F_1$ and $^7F_2$ for Eu^{3+}, $^6H_{7/2}$ for Sm^{3+} – that were thermally populated at room temperature; when these were taken into account (using a Boltzmann factor of $\exp(-\Delta E/kT)$), good agreement was obtained.

Low symmetry crystals or strongly distorted ligand fields cause anisotropic susceptibility behaviour, though this is not manifest when studying powdered samples. Deviations from Curie-type behaviour tend to be noticeable at low temperatures, owing to zero-field splitting effects.

2.7 Physical methods for investigating lanthanide complexes

The range of coordination numbers adopted by the lanthanides and the relative insensitivity of f electrons to their environment means that one cannot use measurements of magnetic moment and electronic spectra to distinguish between coordination numbers in the way that is possible for octahedrally and tetrahedrally coordinated Co^{2+}, for example. These techniques can still yield valuable information though.

2.7.1 Electronic spectra

The $4f$–$4f$ transitions in the electronic spectra of Ln ions can rarely be used diagnostically; however, the octahedral hexahalide complexes LnX_6^{3-}

(X = Cl, Br) have particularly weak extinction coefficients for the absorption bands, an order of magnitude lower than for the aquo-ions, ascribed to the high symmetry of the environment.

Certain transitions are said to be 'hypersensitive' to changes in the symmetry and strength of the ligand field, in that they display shifts of maxima, usually to longer wavelength, band splitting and intensity variation. This is most marked for the $^4I_{9/2} \rightarrow {}^2H_{9/2}, {}^4F_{5/2}$ and $^4I_{9/2}, {}^4G_{5/2}, {}^2G_{7/2}$ transitions in the Nd^{3+} ion, but other ions display the effect, such as Er^{3+} and Ho^{3+}. Figure 2.13 shows how the profile of the band due to the $^4I_{9/2} \rightarrow {}^2H_{9/2}, {}^4F_{5/2}$ transition in the spectrum of $Nd^{3+}(aq)$ resembles that of the 9-coordinate $Nd(OH_2)_9^{3+}$ ions in solid $Nd(BrO_3)_3.9H_2O$ yet is significantly different from 8-coordinate Nd^{3+} ions in the solid sulphate and chloride. The spectrum of Nd^{3+} in concentrated HCl suggests 8-coordination in this solution.

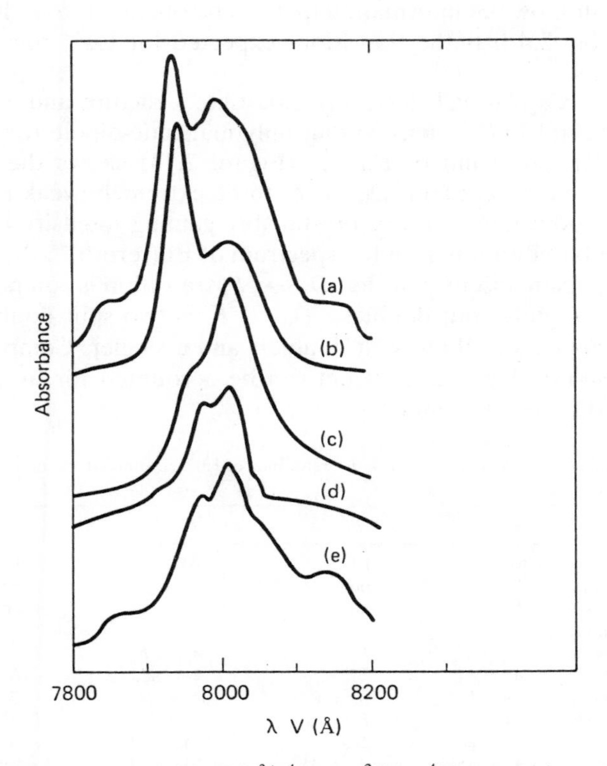

Figure 2.13 Spectra of the Nd^{3+} $^4I_{9/2} \rightarrow {}^2H_{9/2}, {}^4F_{5/2}$ transitions: (a) solid $Nd(BrO_3)_3. 9H_2O$; (b) 5.35×10^{-2} M Nd^{3+} in water; (c) 5.35×10^{-2} M Nd^{3+} in 11.4 M HCl; (d) solid $NdCl_3.6H_2O$; (e) solid $Nd_2(SO_4)_3.8H_2O$. (a) and (b) have 9-coordinate Nd^{3+}, (c)–(e) are 8-coordinate [from D. G. Karraker, *Inorg. Chem.*, **7** (1968) 474]

2.7.2 Luminescence spectra

As outlined in section 2.6.3, luminescence spectra can in principle be achieved for any lanthanide ion not having a f^0, f^7, or f^{14} electron configuration, best results being obtained from Eu^{3+}, Tb^{3+}, Sm^{3+} and Dy^{3+}.

Eu^{3+} is particularly convenient as the $^5D_0$ excited state is non-degenerate and thus cannot be split by a ligand field, nor can the $^7F_0$ ground state, so that if more than one line is seen for the $^5D_0 \rightarrow {}^7F_0$ transition, it shows more than one Eu^{3+} site. This is an electric dipole transition, and is allowed only in low-symmetry environments, particularly when there is no inversion centre. The $^5D_0 \rightarrow {}^7F_1$ transition is magnetic-dipole-allowed and its intensity is little affected by the environment. The $^5D_0 \rightarrow {}^7F_2$ transition is electric-dipole in origin; it is absent if the ion is on an inversion centre and its intensity is again environment-sensitive. Two examples will show the information that can be obtained from fluorescence spectra. Table 2.4 lists the transitions expected for Eu^{3+} ions in various environments.

The salts $Cs_2NaLnCl_6$ have the elpasolite structure and contain perfectly octahedral $LnCl_6^{3-}$ ions so that only magnetic-dipole transitions are expected. The spectrum of $EuCl_6^{3-}$ (Figure 2.14) shows the $^5D_0 \rightarrow {}^7F_0$ transitions to be absent and $^5D_0 \rightarrow {}^7F_2$ to be extremely weak at 77K (it is stronger at room temperature, presumably gaining intensity by coupling with molecular vibrations). In the spectrum of $Eu(terpy)_3^{3+}$, the $^5D_0 \rightarrow {}^7F_0$ transition is again absent, but the $^5D_0 \rightarrow {}^7F_1$ transition is composed of two lines, one a slightly split doublet; $^5D_0 \rightarrow {}^7F_2$ is two split doublets, while $^5D_0 \rightarrow {}^7F_4$ appears as three split doublets and a singlet. Comparison with Table 2.4 shows that the spectrum can be accounted for by assuming a slight distortion of D_3 symmetry.

Table 2.4 Number and degeneracy of $^5D_0$–^7F_J transitions of Eu^{3+} in some of the more common symmetries

	$^7F_0$	$^7F_1$	$^7F_2$	$^7F_3$	$^7F_4$
Symmetry	ED	MD	ED	MD	ED
I_h	none	T_{1g}	none	none	none
O_h	none	T_{1g}	none	T_{1g}	none
T_d	none	T_1	T_2	T_1	T_2
D_{4h}	none	$A_{2g} + E_g$	E_g	$A_{2g} + E_g$	none
D_{4d}	none	$A_2 + E_3$	E_1	$A_2 + E_3$	$B_2 + E_1$
D_{2d}	none	$A_2 + E$	$B_2 + E$	$A_2 + 2E$	$B_2 + 2E$
D_{3h}	none	$A_2' + E''$	E'	$A_2' + E''$	$A_2'' + 2E$
D_3	none	$A_2 + E$	$2E$	$2A_2 + 2E$	$A_2 + 3E$
C_{3v}	A_1	$A_2 + 2E$	$A_1 + 2E$	$2A_2 + 2E$	$2A_1 + 3E$
C_3	A	$A + E$	$A + 2E$	$3A + 2E$	$3A + 3E$
C_{2v}	A_1	$A_2 + B_1 + B_2$	$2A_1 + B_1 + B_2$	$2A_2 + 2B_1 + 2B_2$	$3A_1 + 2B_1 + 2B_2$
C_1	A	$3A$	$5A$	$7A$	$9A$

Fuller tabulations can be found in, for example:
J. H. Forsberg, *Coord. Chem. Revs*, **10** (1973) 205;
K. B. Yatsimirskii and V. K. Davidenko, *Coord. Chem. Revs*, **27** (1979) 223;
J. C. G. Buenzli, in J. C. G. Buenzli and G. R. Choppin (eds), *Lanthanide Probes in Life, Chemical and Earth Sciences*, Elsevier, 1989, 415.

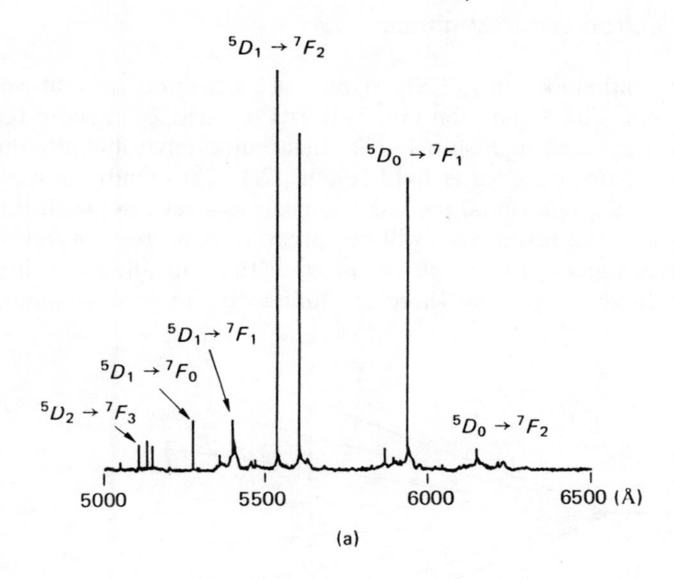

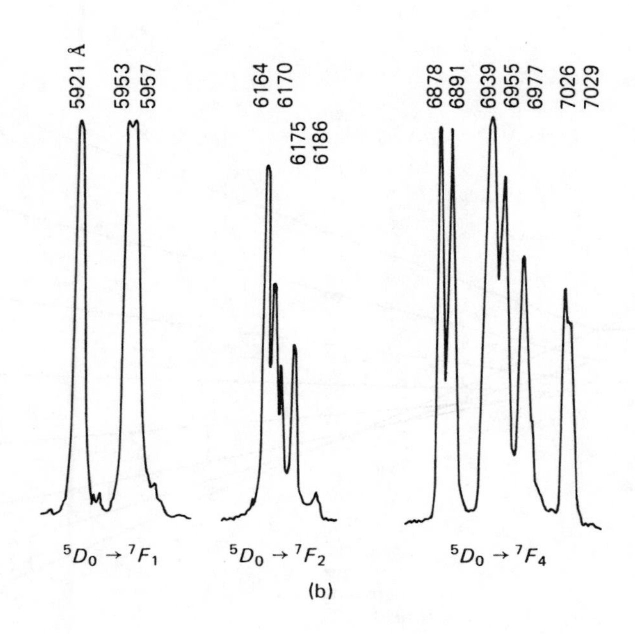

Figure 2.14 Fluorescence spectra of: (a) $EuCl_6^{3-}$ ions in $Cs_2NaEuCl_6$ at 77K; (b) $Eu(terpy)_3^{3+}$ ions in $Eu(terpy)_3(ClO_4)_3$ at 180K [from O. C. Serra *et al.*, *Inorg. Chem.*, **15** (1976) 505, reprinted by permission of the American Chemical Society © 1976; D. A. Durham *et al.*, *J. Inorg. Nucl. Chem.*, **31**(1969) 833, reprinted by permission of Pergamon Press PLC © 1969]

2.7.3 Electron spin resonance

For most lanthanide ions, ESR signals are obtained only at very low temperatures. The *S*-state ion Gd^{3+} affords resonances at room temperature (like Cr^{3+} and high-spin Fe^{3+}). In a cubic environment, the resonances all occur at the same field (Figure 2.15); normally, however, the distorted crystal field round the Gd^{3+} ion causes a zero-field splitting, with the result that the resonances will be spread over a range of fields. In an 'axial' environment, there will be 14 transitions in all, more in a low-symmetry field. At present there are limited data on coordination com-

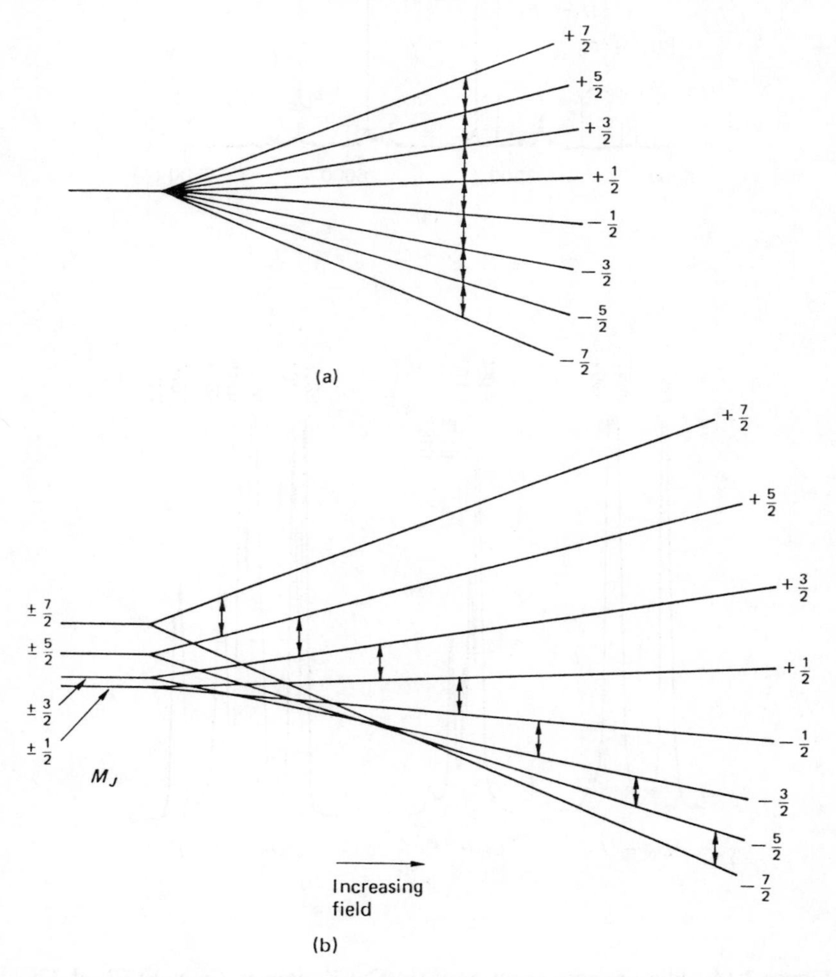

Figure 2.15 Energy levels of the $^8S_{7/2}$ ground state of Gd^{3+} in a magnetic field: (a) no zero-field splitting, all ESR transitions occur at same field; (b) zero-field splitting

pounds, though Gd^{3+} has also been used as a probe for a number of proteins. One interesting result has been obtained for the three-coordinate compound $Gd[N(SiMe_3)_2]_3$, with a very strong axial field, which exhibits the effective g values $g_\perp = 8$, $g_\parallel = 2$, a striking analogy with the corresponding Fe^{3+} ($g_\perp = 6$, $g_\parallel = 2$) and Cr^{3+} ($g_\perp = 4$, $g_\parallel = 2$) compounds.

2.7.4 Mossbauer spectroscopy

Several lanthanides have isotopes amenable to study; most data have been reported for ^{151}Eu. Spectra tend to be single lines, with characteristic line-widths of around 3 mm s^{-1}; the limited studies made thus far of coordination compounds of Eu^{3+} have been relatively uninformative, the largely ionic bonding ensuring that the isomer shift is rather insensitive to ligand. However, isomer shift data can clearly identify the difference between Eu^{2+} and Eu^{3+}, as there is a difference of some 12 mm s^{-1} between the oxidation states. Thus in the compound Eu_3S_4, separate Eu^{2+} and Eu^{3+} resonances are seen at low temperatures.

2.7.5 Magnetic susceptibility measurements

As indicated in section 2.6.6, the magnetic moment of a given lanthanide ion is essentially independent of its environment. The moment does depend on oxidation state and can thus be used to distinguish between Eu^{3+} ($\mu_{eff} = 3.5\ \mu_B$) and Eu^{2+} ($\mu_{eff} = 7.9\mu_B$); Ce^{3+} ($\mu_{eff} = 2.5\mu_B$) and Ce^{4+} (diamagnetic); Yb^{3+} ($\mu_{eff} = 4.5\mu_B$) and Yb^{2+} (diamagnetic), among others.

2.7.6 NMR spectra

Section 2.11 discusses applications of lanthanides in NMR. Some studies, however, have been concerned with the geometry of the lanthanide complex; for example, all the lanthanides form complexes $Ln(glycollate)_3(H_2O)_3$ in solution. Analysis of the lanthanide-induced shifts in the ^1H, ^{13}C and ^{17}O NMR spectra of the complexes of all 14 lanthanides has allowed assignment of the positions of all the atoms in the complexes.

2.7.7 Infra-red spectroscopy

This technique is, of course, routinely employed in inorganic chemistry. It is of particular value in lanthanide chemistry in detecting the coordination – or lack of it – of anions like nitrate and perchlorate; this may well vary as coordination numbers change across a series of apparently identical complexes.

Thus $Yb\,en_4(NO_3)_3$ is shown to contain only ionic nitrate, but $La\,en_4(NO_3)_3$ has both ionic and coordinated nitrate, while only coordinated nitrate is found in $La\,bipy_2(NO_3)_3$ (confirmed by X-ray diffraction).

2.7.8 Other techniques

Among other techniques are a number that may be applied to study the stoichiometry of complex formation in solution, such as Job's plots, as well as enthalpometric and conductometric titrations. An example of conductometric titration consists of studying the reaction between the tridentate ligand diethylene-triamine (dien) and neodymium nitrate:

$$Nd(NO_3)_3 \longrightarrow Nd(dien)(NO_3)_3 \longrightarrow Nd(dien)_2(NO_3)_3$$

Adding the ligand solution gradually, it was found that the complex formed with a 1:1 ligand: neodymium ratio was essentially a non-electrolyte, probably a 9-coordinate $Nd\,dien(NO_3)_3$ species, containing bidentate nitrates. As the second molecule of ligand was added, the conductivity rapidly increased, suggesting an ionic formula, such as 10-coordinate $Nd(dien)_2(NO_3)_2^+\ NO_3^-$.

2.7.9 Lanthanides as bio-probes

Over the last two decades, there has been increasing use of lanthanide ions as spectroscopic probes for biological systems, particularly as a substitute for Ca^{2+}, which is not spectroscopically active. Lanthanides have no established biological role but have low toxicity.

Apart from being a similar size to Ca^{2+}, the lanthanides also have the facility of essentially non-directional electrostatic bonding, so that they are fairly easily accommodated without distorting binding sites. They are known to substitute isomorphously for calcium in some metalloproteins (parvalbumin, thermolysin) and to activate certain Ca-based systems in the absence of calcium.

Obviously the weak electronic spectra of lanthanides are not a good 'probe', but luminescence spectra of Eu^{3+} and Tb^{3+} are being extensively used. Gd^{3+} has been substituted as an EPR probe for Ca in parvalbumin and for Fe^{3+} in transferrin. Lanthanide ions have found wide acceptance as an aid to NMR spectral assignment (see section 2.11).

2.8 Lanthanide binary compounds

2.8.1 Lanthanide halides

Trihalides

These have been characterised for all the lanthanides (with the possible exception of EuI_3) and afford an excellent vehicle for studying the effect upon structure of varying the radii of cation and anion. They are important starting materials in the synthesis of lanthanide compounds; all except the fluorides are deliquescent.

The trifluorides are insoluble in water and come out of solution as hydrates. The anhydrous fluorides can be prepared by dehydration of this hydrate or by hydrofluorination direct from the oxide:

$$LnF_3.\tfrac{1}{2}H_2O \xrightarrow[\text{or } 600°/\text{HF}]{300°,\ in\ vacuo} LnF_3$$

$$Ln_2O_3 \xrightarrow[700°]{HF(g)} LnF_3$$

$$Ln_2O_3 \xrightarrow[300°]{NH_4HF_2} NH_4LnF_4 \xrightarrow{450°} LnF_3 + NH_4F$$

Dehydration of the other halides tends to lead to oxyhalide formation but there are plenty of routes for chlorides:

$$Ln_2O_3 \xrightarrow[300°]{CCl_4} LnCl_3$$

$$Ln_2O_3 \xrightarrow[\text{heat}]{\text{exc. } NH_4Cl} LnCl_3$$

$$LnCl_3.xH_2O \xrightarrow[105\text{–}350°]{HCl(g)} LnCl_3$$

$$LnCl_3.xH_2O \xrightarrow[\text{propane}]{2,2'\text{-dimethoxy-}} LnCl_3.xMeOH \xrightarrow[150°]{in\ vacuo} LnCl_3$$

$$LnCl_3.xH_2O \xrightarrow[\text{reflux}]{SOCl_2} LnCl_3$$

The recommended method for bromides is the ammonium bromide route, also applicable to chlorides and iodides:

$$Ln_2O_3 \xrightarrow[200\text{–}300°]{\text{exc. } NH_4Br} (NH_4)_2LnBr_5,\ (NH_4)_3LnBr_6 \xrightarrow[in\ vacuo]{350\text{–}400°} LnBr_3$$

$$LnCl_3 \xrightarrow[400\text{–}600°]{HBr(g)} LnBr_3$$

$$LnBr_3.6H_2O \xrightarrow[100°]{\textit{in vacuo}} LnBr_3 \text{ (Gd – Lu)}$$

$$LnCl_3 \xrightarrow[\text{heat}]{HI/H_2} LnI_3$$

$$LnI_3.xH_2O \xrightarrow[\text{heat}]{\text{exc. } NH_4I} LnI_3$$

$$Ln + I_2 \xrightarrow{800°} LnI_3$$

$$La \xrightarrow[330°]{HgI_2} LaI_3$$

The very hygroscopic halides should be purified by sublimation, though in view of possible oxyhalide formation, contact with hot silica is undesirable.

For the early trifluorides, the structure adopted (Table 2.5) is that of LaF_3 ('tysonite') in which the metal ion has nine near-neighbour fluorides (tricapped trigonal prism) but with two more a little further away, in what might be termed 9 + 2 coordination. Later lanthanides have the YF_3 structure; here there are 8 fluorides about 2.3 Å away and a ninth at 2.6 Å, in a rather distorted tricapped trigonal prism (Figure 2.16).

The early trichlorides have the 9-coordinate UCl_3 structure, $TbCl_3$ that of $PuBr_3$ (8-coordinate) and subsequent elements the 6-coordinate $AlCl_3$ structure; the bromides display a similar sequence of 9 (La–Pr), 8 (Nd–Eu) and 6-coordination (Gd–Lu). The iodides are 8-coordinate (La–Pm) or 6-coordinate (Sm–Lu). There are frequently high-temperature phases adopting another structure form, for example, PmI_3 also adopts the 6-coordinate $FeCl_3$ structure.

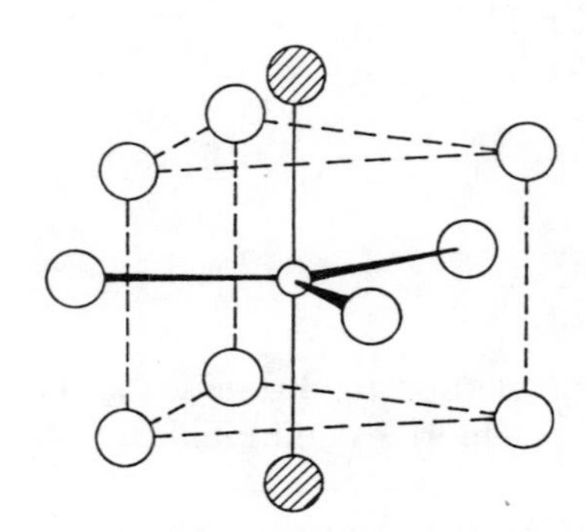

Figure 2.16 The LaF_3 structure, showing the two distant fluorines, completing the (9 + 2)-coordination (shaded). In the YF_3 structure, yttrium is bound to 9 fluorines in an approximately trigonal prismatic array

Table 2.5 Structures of the lanthanide trihalides

	F	Cl	Br	I
La	LaF_3	UCl_3	UCl_3	$PuBr_3$
Ce	LaF_3	UCl_3	UCl_3	$PuBr_3$
Pr	LaF_3	UCl_3	UCl_3	$PuBr_3$
Nd	LaF_3	UCl_3	$PuBr_3$	$PuBr_3$
Pm	LaF_3	UCl_3	$PuBr_3$	$PuBr_3$
Sm	YF_3	UCl_3	$PuBr_3$	$FeCl_3$
Eu	YF_3	UCl_3	$PuBr_3$	$FeCl_3$(?)
Gd	YF_3	UCl_3	$FeCl_3$	$FeCl_3$
Tb	YF_3	$PuBr_3$	$FeCl_3$	$FeCl_3$
Dy	YF_3	$AlCl_3$	$FeCl_3$	$FeCl_3$
Ho	YF_3	$AlCl_3$	$FeCl_3$	$FeCl_3$
Er	YF_3	$AlCl_3$	$FeCl_3$	$FeCl_3$
Tm	YF_3	$AlCl_3$	$FeCl_3$	$FeCl_3$
Yb	YF_3	$AlCl_3$	$FeCl_3$	$FeCl_3$
Lu	YF_3	$AlCl_3$	$FeCl_3$	$FeCl_3$
Y	YF_3	$AlCl_3$	$FeCl_3$	$FeCl_3$

Table 2.6 MX_2 (X = F, Cl, Br, I) compounds formed by the lanthanides

	F	Cl	Br	I
La				*[9]
Ce				*[9]
Pr				*[9]
Nd		√[2]	√[2]	√[3]
Pm				
Sm	√[1]	√[2]	√[2, 3]	√[6]
Eu	√[1]	√[2]	√[3]	√[6]
Gd				*[9]
Tb				
Dy		√[3]	√[4]	√[7]
Ho				
Er				
Tm		√[4]	√[4]	√[8]
Yb	√[1]	√[4]	√[4, 5]	√[8]
Lu				

[1] CaF_2 type, CN = 8.
[2] $PbCl_2$ type, CN = 7 + 2.
[3] $SrBr_2$ type, CN = 7, 8.
[4] SrI_2 type, CN = 7.
[5] $CaCl_2$ type, CN = 6.
[6] EuI_2 type, CN = 7.
[7] $CdCl_2$ type, CN = 6.
[8] CdI_2 type, CN = 6.
[9] $MoSi_2$ type, CN = 8.
* 'Metallic' di-iodide.

The series exhibits a trend (also noted in the dihalides) to decreasing coordination number as the radius of Ln^{3+} decreases or as the radius of the halide ion increases. Consider, say, the chlorides with the UCl_3 structure; as the Ln^{3+} ion gets smaller, the chloride ligands will be drawn closer until a stage is reached where non-bonding Cl–Cl repulsive forces make the 8-coordinate structure energetically more favourable.

Dihalides

These are shown in Table 2.6. They fall into two classes, normal 'salt-like', and 'metallic'. This latter class comprises the di-iodides of La, Ce, Pr and Gd, which have a metallic lustre and high conductivity. They are very good reducing agents, reacting immediately with air and water, and are believed to have the structure $Ln^{3+}(I^-)_2(e^-)$ with the odd electron delocalised in a conduction band. They are generally made by reproportionation:

$$La + 2LaI_3 \longrightarrow 3LaI_2$$

The 'salt-like' dihalides have magnetic and spectroscopic properties consistent with $Ln^{2+}(X^-)_2$ structures, for example the compounds are non-conductors. They are prepared by various methods, such as hydrogen reduction and reproportionation:

$$MX_3 \xrightarrow[\text{heat}]{H_2} MX_2$$

(M = Eu, Yb, X = Cl, Br, I; M = Eu, X = F)

$$NdCl_3 \xrightarrow[\text{heat}]{Nd} NdCl_2$$

$$Tm \xrightarrow[\text{heat}]{HgI_2} TmI_2$$

$$M \xrightarrow[\text{(ii) warm}]{\text{(i) } NH_4X/NH_3(l)} MX_2$$

(M = Eu, Yb; X = Cl, Br, I)

Only those of europium are stable in aqueous solution. Such halides may also be made by reductive methods in solution as solvates (see section 2.12.4). Additionally, the M^{2+} ions can be doped into CaF_2 lattices (this has been done principally to study the electronic spectra).

The halides demonstrate the expected trends to decreasing coordination number with increasing radius of the anion and decreasing size of the cation. Promethium would also be expected to form dihalides (except for the fluoride) but its radioactivity is a bar to study.

Other reduced halides

A number of halides with non-integral stoichiometrics have been made. Many of these contain both M^{2+} and M^{3+} but there are a number of interesting clusters, such as $GdCl_{1.5}$, which has a chain structure based on octahedra of gadolinium atoms sharing trans-edges. Other halides like GdCl and TbCl have layer structures.

High oxidation states

The only halides in this class are MF_4 (M = Ce, Pr, Tb) discussed in section 2.12.1.

2.8.2 Oxides

These can be made by heating lanthanide oxy-compounds such as the nitrate, carbonate or hydroxide: for example

$$4Ln(NO_3)_3 \longrightarrow 2Ln_2O_3 + 12NO_2 + 3O_2$$

The oxides of those metals with a tendency to the +4 state need a reducing atmosphere to prepare M_2O_3 (Ce, Pr, Tb); otherwise the higher oxides like CeO_2 may be reduced with hydrogen.

Three types of structure are adopted. In the *A*-type, adopted by La_2O_3–Sm_2O_3, there is capped octahedral 7-coordination. The *B*-type (Pr–Gd, stabler at high temperatures) has three sites, one a distorted capped octahedron approximating to 6-coordination, the other two face capped trigonal prismatic 7-coordination. In the *C*-type structure, 6-coordination is formed, but severely distorted from octahedral in most sites.

The oxides dissolve in acid to form salts but show no amphoteric tendencies; they tend to absorb CO_2 and H_2O from the atmosphere.

Europium(II) oxide is the best-defined lower oxide, prepared by reduction of Eu_2O_3 with Eu, La or C:

$$Eu + Eu_2O_3 \xrightarrow{800\text{–}2000^\circ} 3EuO$$

Attempts to produce similar oxides of other divalent metals led to compounds believed to be $SmN_{1/2}O_{1/2}$ and $YbC_{1/2}O_{1/2}$. Use of high pressures as well as high temperatures yields the golden NdO. SmO and YbO have been obtained similarly.

$$Nd + Nd_2O_3 \xrightarrow[1000^\circ]{50\text{ kbar}} 3NdO$$

Mixed valence oxides M_3O_4 (M, for example, Eu or Sm), which are $M^{2+}(M^{3+})_2(O^{2-})_4$, also exist.

2.8.3 Hydroxides

These are precipitated by the addition of alkali to solutions of lanthanide salts. Hydrothermal crystallisation at high temperature and pressure affords the crystalline hydroxides $Ln(OH)_3$ (under different conditions, LnO(OH) obtains). These compounds are basic enough to react with atmospheric carbon dioxide to afford the carbonates and are, of course, dissolved by acids.

With the exception of $Lu(OH)_3$ (which has the 6-coordinate $In(OH)_3$ structure) all the lanthanide hydroxides adopt the 9-coordinate UCl_3 structure.

2.8.4 Oxide superconductors

The report in late 1986 by Bednorz and Muller of a superconducting transition temperature of 38K in $La_{1.8}Sr_{0.2}CuO_x$ marked a major breakthrough after years in which the highest T_c recorded was 23K. In March, 1987 Wu, Chu *et al.* reported that the compound $YBa_2Cu_3O_{7-\delta}$ $(0 \leqslant \delta \leqslant 1)$ exhibited a T_c of 92K (zero resistance is found at 90K, above the boiling point of nitrogen). The synthesis of such a material has obvious implications in areas such as magnetic devices, power transmission and rapid communications.

The structure of $YBa_2Cu_3O_7$ is shown in Figure 2.17; it has an oxygen-deficient layered perovskite structure and formally contains Cu^{2+} and Cu^{3+}. Removing oxygen atoms from the O(1) sites leads to the formation of semiconductor $YBa_2Cu_3O_6$. In the former, the copper ions have square-pyramidal and square-planar coordination. The properties of the materials depend critically on factors such as annealing temperature and quenching rate: vacuum annealing removes oxygens from the Cu(1) plane reversibly but not from the Cu(2) plane.

Current theories of superconductivity suggest that an electron travelling through such a lattice distorts it so that a following electron senses the distortion and follows the first one, a so-called 'Cooper-pair'.

2.8.5 Sulphides

These exist in a range of stoichiometries – MS, M_2S_3, M_3S_4, MS_2. The most important are the sesquisulphides M_2S_3, made either by direct combination or by:

$$2LnCl_3 + 3H_2S \xrightarrow{\text{heat}} Ln_2S_3 + 6HCl \text{ (except Eu)}$$

$$4EuS_2 \xrightarrow{\text{heat}} 2Eu_2S_3 + 2S$$

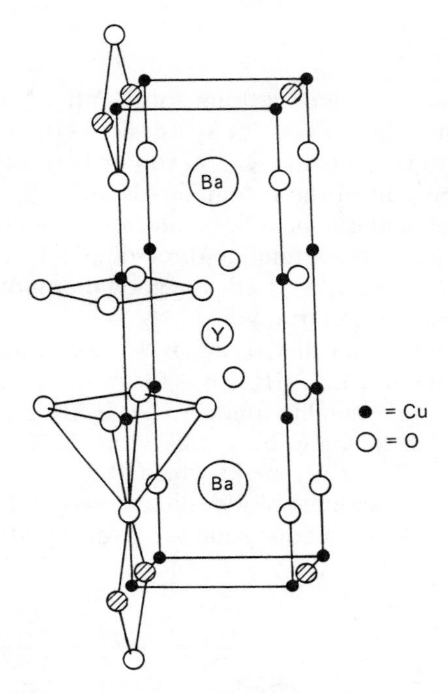

Figure 2.17 The structure of $YBa_2Cu_3O_7$ (the shaded atoms are those removed to create the oxygen-deficient phases up to $YBa_2Cu_3O_6$)

These are genuine M^{3+} compounds and exhibit increasing coordination numbers as the size of the lanthanide ion increases – Lu_2S_3 and Yb_2S_3 have the corundum structure (CN = 6), Dy_2S_3–Tm_2S_3 the Ho_2S_3 structure (CN = 6, 7) and La_2S_3–Dy_2S_3 (except Eu) the Gd_2S_3 structure (CN = 7, 8). There is also the Ce_2S_3 structure (M = La–Sm) which is based on Th_3P_4 with cation vacancies, having 8-coordination. The sulphides M_3S_4 (obtained by loss of sulphur on heating M_2S_3) are formed by the metals La–Eu and also have the Th_3P_4 structure. Some of these like Ce_3S_4 are metallic conductors and thus probably $(Ce^{3+})_3(S^{2-})_4(e^-)$ while the semiconductor Sm_3S_4 and the Eu analogue are probably $M^{2+}(M^{3+})_2(S^{2-})_4$.

The monosulphides MS, formed by direct reaction at 1000°, adopt the NaCl structure but again exhibit more than one type of bonding. EuS and YbS are probably $M^{2+}S^{2-}$. Others like the golden LaS exhibit metallic properties; CeS has the magnetic moment expected for Ce^{3+} – thus they appear to be $M^{3+}S^{2-}(e^-)$.

SmS is normally obtained as a black semiconductor phase, which can be represented $Sm^{2+}S^{2-}$. It can be turned into a golden metallic phase by the action of pressure, mixing with other monosulphides, or more remarkably by polishing pressed samples or scratching single crystals. Considerable research is being undertaken to understand this phenomenon.

2.8.6 Borides

A variety of these exist: thus, yttrium forms YB_2, YB_4, YB_6, YB_{12} and YB_{66}. The lanthanide borides can be synthesised either by direct synthesis at 2000° or by heating the oxide with boron or boron carbide at 1800°.

The most important of these compounds are MB_6: YB_6 is unaffected by hot acid or alkali and is a metallic conductor; like other MB_6 it contains linked B_6 octahedra in a continuous three-dimensional framework, with inserted Y^{3+} ions (Figure 2.18); LaB_6 is a good thermionic emitter, used in the thermal cathodes of electron guns.

It has been pointed out that B_6 needs two electrons to fill its bonding MOs, so that a structure $Ln^{3+}(B_6^{2-})(e^-)$ for most lanthanide hexaborides would account for their metallic conductivity. EuB_6 and YbB_6 have higher resistivities and are believed to be salt-like $Ln^{2+}B_6^{2-}$.

YB_{12} has the UB_{12} structure in which B_{12} cuboctahedra and metal atoms are packed in a sodium-chloride-like structure. Lower borides tend to decompose to MB_6 on heating, and are hydrolysed by water to boron hydrides.

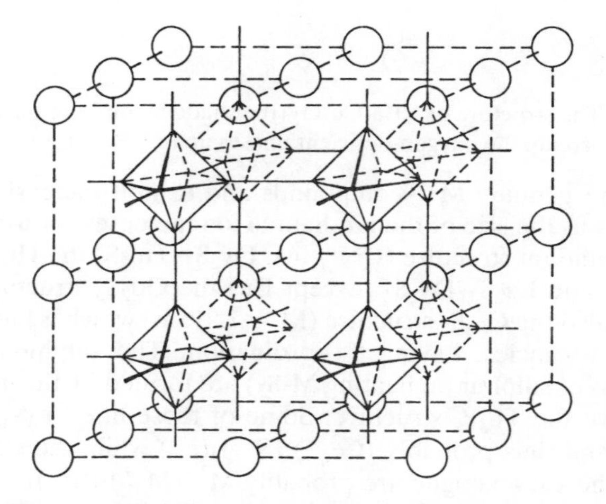

Figure 2.18 YB_6 structure with linked B_6 octahedra

2.8.7 Carbides

The most familiar of these compounds are the dicarbides LnC_2, made by heating the metal oxide, hydride or the metal itself with carbon. They adopt the CaC_2 structure (C–C 1.28–1.29 Å) but are metallic conductors: they may thus be regarded as $Ln^{3+}(C_2^{2-})(e^-)$. Similarly Ln_2C_3 have the Pu_2C_3 structure, where C_2 units are again present; other compounds known are LnC, Ln_2C and Ln_3C.

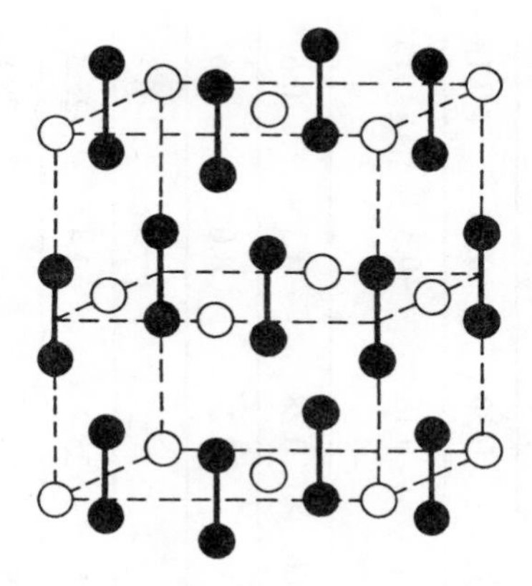

Figure 2.19 CaC_2 structure of LnC_2 compounds

The carbides are readily hydrolysed, particularly under acidic conditions, affording a complex mixture of hydrocarbons, mainly C_2H_2, whose composition varies with temperature.

2.8.8 Pnictides

Compounds MQ (Q = N, P, As, Sb, Bi) are formed, usually by direct synthesis from the elements at 1000°. The nitrides can also be made from the lanthanide hydride by reaction with nitrogen or ammonia. They have the sodium chloride structure and are hydrolysed to the non-metal hydride QH_3.

2.9 Stability constants of lanthanide complexes

A stability (formation) constant, K_1, can be written for the complexing reaction between a lanthanide ion and a ligand L^{n-}:

$$M^{3+}(aq) + L^{n-}(aq) \longrightarrow ML^{(3-n)+}(aq)$$

$$K = \frac{[ML^{(3-n)+}(aq)]}{[M^{3+}(aq)][L^{n-}(aq)]}$$

Table 2.7 shows the stability constants for a range of 1:1 complexes formed between lanthanum and lutetium with a range of ligands, together with comparative data for several other metals.

The values of K_1 for lutetium are consistently greater than those for lanthanum, as expected on electrostatic grounds for a smaller ion with greater charge density; similarly the very small Sc^{3+} ion gives even higher values, as do Fe^{3+}, Th^{4+} and U^{4+}. On the other hand, Ca^{2+}, an ion of

Table 2.7 Aqueous stability constants (log K_1) for lanthanide (3+) and other ions at 25°

	La^{3+}	Lu^{3+}	Y^{3+}	Sc^{3+}	Fe^{3+}	Cu^{2+}	Ca^{2+}	U^{4+}	UO_2^{2+}	Th^{4+}
F^- (1M)	2.67	3.61	3.60	6.2*	5.2	0.9	0.6	7.78	4.54	7.46
Cl^- (1M)	−0.1	−0.4	−0.1	0	0.63	−0.06	−0.11	0.30	−0.10	0.18
Br^- (1M)	−0.2 (Eu^{3+})	—	−0.15	−0.07	−0.2	−0.5*	—	0.18	−0.3	−0.13*
NO_3^- (1M)	0.1	−0.2	—	0.3	−0.5	−0.01	−0.06	0.28	−0.3	0.67*
OH^- (0.5M)	4.7	5.8	5.4	9.3*	11.0*	6.3*	1.3*	12.2	8.0*	9.6*
$acac^-$ (0.1M)	4.94	6.15	5.89	8	10	8.16	—	—	—	8
$EDTA^{4-}$ (0.1M)	15.46	19.8	18.1	23.1	25.0	18.7	10.6	25.7	7.4	25.3
$DTPA^{5-}$ (0.1M)	19.5	22.4	22.1	24.4	28	21.4	10.8	—	—	28.8
CH_3COO^- (0.1M)	1.82	1.85	1.68	—	3.38	1.83	0.5*	—	2.61	3.89*

*Data for solutions of slightly different ionic strength.

Data selected from A. E. Martell and R. M. Smith, *Critical Stability Constants*, Vols 1–5, Plenum Press, New York, 1974 onwards.

rather similar size to the lanthanides but with a smaller charge, forms weaker complexes.

The lanthanides form stabler complexes with fluoride than with chloride and bromide, as expected for a metal ion with a high charge:radius ratio, a 'hard' acid. Polydentate ligands form particularly stable complexes (the 'chelate' effect); thermodynamic data show that the main driving force for this is the favourable entropy change:

$$La(H_2O)_9^{3+}(aq) + EDTA^{4-}(aq) \longrightarrow La(EDTA)(H_2O)_3^-(aq) + 6H_2O(l)$$

$$\Delta G = -87\ \text{kJ mol}^{-1};\ \Delta H = -12\ \text{kJ mol}^{-1};\ \Delta S = 251\ \text{J K}^{-1}\ \text{mol}^{-1}$$

Similar results are obtained for other ligands, ΔH having small exothermic or endothermic values.

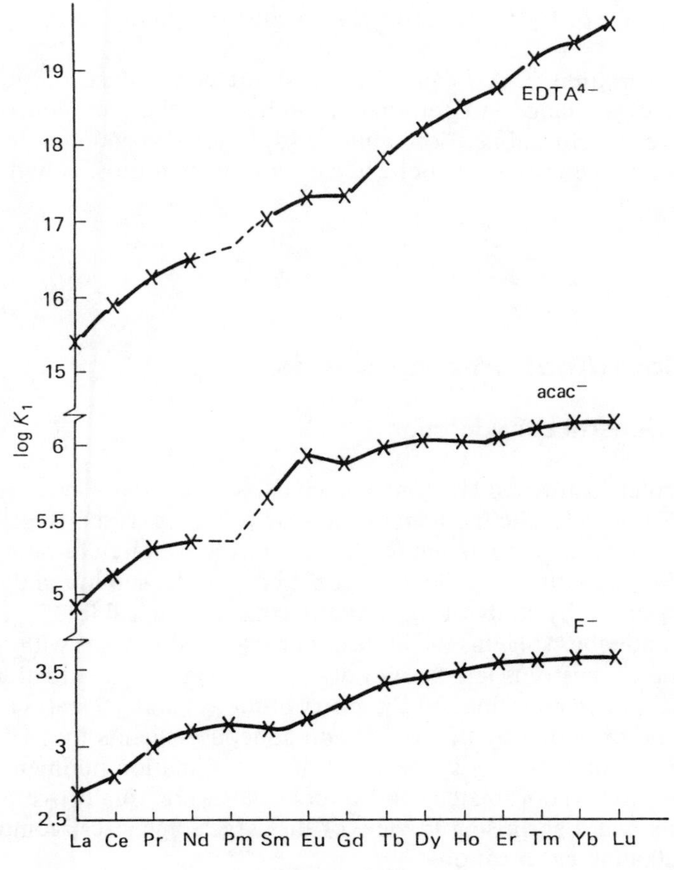

Figure 2.20

Figure 2.20 shows how log K_1 varies for representative complexes across the lanthanide series – while there is, as already mentioned, a general trend to increasing complex stability, there are some inflections in the curve, particularly in mid-series, the 'gadolinium break'. Such effects in the case of transition metals are usually attributed to ligand-field effects, but as these amount to 100–200 cm^{-1} in the case of lanthanides, and since the inflections vary in position, they are discounted here. It is believed that there is a decrease in coordination number of the aquo-ion; this would result in a decrease in ΔS, which would manifest itself in a decrease in K:

$$\Delta G = -RT \ln K = \Delta H - T\Delta S$$

In the reactions

$$LnX_3.9H_2O(s) + 3L^-(aq) \longrightarrow LnL_3(aq) + 3X^-(aq) + 9H_2O(l)$$

$$X = BrO_3 \text{ or } C_2H_5OSO_3 \text{ (ethyl sulphate)} = 2, 6\text{-dipicolinate}$$

a smooth variation in ΔH (and K) across the series is seen, since any possibility of a change in coordination number of the aquo-ion is eliminated. A very slight stabilisation (about 2 kJ) is noted at gadolinium, which may be a genuine crystal-field effect, far weaker than those noted for the transition metals.

2.10 Coordination compounds

2.10.1 General considerations

The particular features of lanthanide chemistry have already been summarised in section 2.4. The lanthanides may be classified as 'hard' acids and therefore prefer to bind to 'hard' bases – oxygen and fluorine – rather than 'soft' bases with atoms like nitrogen, phosphorus, sulphur and iodine. This is borne out by study of the stability constants in Table 2.7. Though use of non-aqueous solvents facilitates synthesis of complexes with the soft donors, the chemistry is largely that of the most stable +3 oxidation state though the limited coordination chemistry of the +4 and +2 states (section 2.12) can be extended by the use of non-aqueous solvents too. Until the 1960s, it was not generally appreciated that coordination numbers of 8–9 were the norm (as coordinated solvent molecules were usually overlooked) though this is not surprising in view of the large ionic sizes, comparable with the alkaline earth cations.

2.10.2 Complexes of oxygen-donor ligands

Aquo-complexes

The lanthanides form a wide range of hydrated salts, some containing exclusively water molecules in the coordination sphere. Thus $LnX_3.9H_2O$ ($X = BrO_3$, $EtSO_4$, CF_3SO_3) contain $Ln(H_2O)_9^{3+}$ ions where there is tricapped trigonal prismatic coordination of the metal; this has been established by X-ray diffraction studies for several complexes, with the expected contraction in the Ln–O distance as the series is traversed. Curiously, the complexes $Ln(ClO_4)_3.6H_2O$ contain octahedrally coordinated lanthanide ions (Ln = Lu, Tb, Er from diffraction data).

A variety of hydrated salts exist which contain coordinated anions. Thus $LnCl_3.6H_2O$ contain $LnCl_2(OH_2)_6^+$ ions (8-coordinate square antiprism) while $LnCl_3.7H_2O$ have dimeric units $(H_2O)_7LnCl_2Ln(OH_2)_7^{4+}$ (Ln = La–Pr) with 9-coordinate metals. $LnBr_3.6H_2O$ are believed to have the same structure as the corresponding chlorides, but the structure of the iodides $LnI_3.nH_2O$ (Ln = La–Gd, $n = 9$; Tb–Lu, $n = 8$) is unknown, though it is probably $Ln(OH_2)_n^{3+}(I^-)_3$.

The thiocyanates $Ln(NCS)_3(H_2O)_n$ (Ln = La–Nd, $n = 7$; Sm–Lu, Y, $n = 6$) have respectively 9-coordinate (3*N*, 6*O*) and 8-coordinate (3*N*, 5*O*) lanthanides, one water molecule in each being uncoordinated. Similar contraction is seen in the nitrates $M(NO_3)_3.xH_2O$ (M = La–Ce, $n = 6$; M = Pr–Lu, $n = 5$); again one water molecule is uncoordinated so they have 11 and 10 coordination respectively, with bidentate nitrates, X-ray diffraction has established the coordination spheres to be $La(NO_3)_3(H_2O)_5$ and $Pr(NO_3)_3(H_2O)_4$ for specific examples.

The sulphates are complicated. $La_2(SO_4)_3.9H_2O$ and the cerium analogue have two kinds of coordination geometry: one lanthanide is bound to six bidentate sulphates (CN = 12), the other to six water molecules and three monodentate sulphates (CN = 9). There are two series of octahydrates (Ln = Ce, La; Pr–Lu, Y which have 9 and 8-coordination respectively); while in $Ln_2(SO_4)_3.5H_2O$ (La–Nd), the cerium and neodymium ions have been shown to be 9-coordinate (distorted tricapped trigonal prism) by two water molecules and seven monodentate sulphates.

Aquo-complexes in solution

There is considerable evidence to support the view that the coordination number of lanthanide ions is about 9. The electronic absorption spectrum of solutions of neodymium bromate is very similar to that of the $Nd(H_2O)_9^{3+}$ ion in the solid salt, while X-ray and neutron-diffraction studies on solutions suggest a coordination number of 8.5–8.9 for $Nd^{3+}(aq)$. Luminescence life-time studies on europium and terbium salts indicate coordination numbers of 9.6 and 9.0 respectively, giving some weight to the opinion that there is a decrease in coordination number across the series.

Other complexes of O-donor ligands

Like water, with its built-in dipole, other ligands like phosphine oxides and amine *N*-oxides form stable complexes with lanthanides, in non-aqueous solvents.

Thus C_5H_5NO, pyridine *N*-oxide, forms 8-coordinate complexes $Ln(C_5H_5NO)_8(ClO_4)_3$, where the coordination geometry is square antiprismatic in the neodymium compound, distorted towards cubic in the lanthanum analogue (Figure 2.21).

Dimethylsulphoxide complexes ($Ln(DMSO)_8(ClO_4)_3$) can similarly be prepared; nitrate complexes have been extensively studied by X-ray diffraction, with 10-coordination established for $Ln(DMSO)_4(NO_3)_3$ (Ln = La, Pr, Nd, Sm, Eu) and 9-coordination in $Ln(DMSO)_3(NO_3)_3$ (Ln = Dy, Er, Yb, Lu). The change from 10 to 9-coordination seems to occur between gadolinium and terbium. 9-coordination is also found in the tetramethylurea complex $Eu[(Me_2N)_2CO]_3(NO_3)_3$.

Phosphine oxides R_3PO like triphenylphosphine oxide (R = Ph) and hexamethylphosphoramide (HMPA; R = Me_2N) are sterically more demanding. Thus 6-coordination is found in $[Ln(HMPA)_6](ClO_4)_3$ and mer-$Ln(HMPA)_3Cl_3$, the latter an unusual case of complexes of the same stoichiometry being found for scandium, yttrium and all the lanthanides

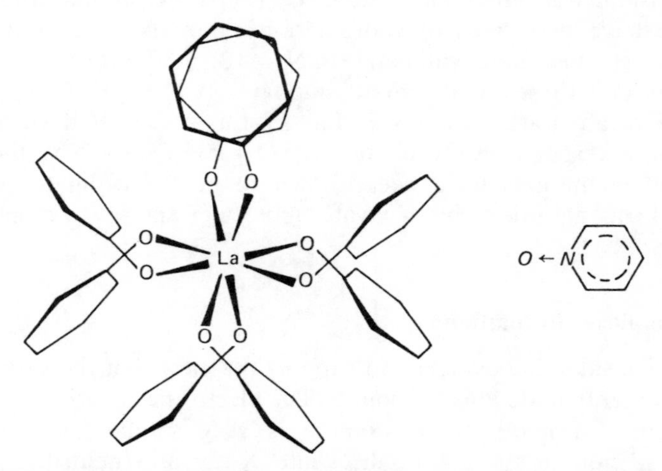

Figure 2.21 The $La(C_5H_5NO)_8^{3+}$ ion; C_5H_5NO is pyridine *N*-oxide [after A. R. Al-Karaghouli and J. S. Wood, *Chem. Commun.*, (1972) 516; reprinted with permission of the Royal Society of Chemistry]

(they have been found to catalyse olefin polymerisation and arene oxidation). An unusual 'one pot' synthetic route has recently been described:

$$\mathrm{La} \xrightarrow[\text{toluene, 90°}]{\mathrm{NH_4X/HMPA}} \mathrm{La(HMPA)}_n\mathrm{X}_3$$

(X = Br, NCS, $y = 4$; X = NO_3, $y = 3$)

The nitrate $La(HMPA)_3(NO_3)_3$ is 9-coordinate (bidentate nitrates), the thiocyanate $La(HMPA)_4X_3$ is 7-coordinate, and the bromide is $[LaBr_2(HMPA)_4]^+Br^-$.

A wide range of triphenyl phosphine oxide and arsine oxide complexes has been made, but there are few structural data. The series $M(NO_3)_3(Ph_3QO)_3$ (Q = P, As) are probably 9-coordinate, while $M(NO_3)_3(Ph_3AsO)_4$ are thought to be the 8-coordinated $[M(NO_3)_2(Ph_3AsO)_4]^+NO_3^-$. $SmCl_3(Ph_3PO)_4$ may be 7-coordinate; Chinese workers have recently reported $La(Ph_3PO)_5Cl^{2+}(FeCl_4^-)_2$ and $[Gd(Ph_3PO)_4Cl_2]^+CuCl_3^-$, both involving 6-coordination.

The alkoxides

There have been many recent developments in this area. Moisture-sensitive compounds $Ln(OR)_3$ have been known for some 30 years; molecular weight measurements suggested oligomeric structures. This may well be an over-simplification – attempted synthesis of $Ln(OPr^i)_3$ (Ln = Sc, Y, Yb) afforded the pentanuclear $Ln_5O(OPr^i)_{13}$ (Figure 2.22) in which all the lanthanide atoms are 6-coordinate. Under different conditions, chloride-containing species like $Nd_6Cl(OPr^i)_{17}$ and $Y_3(OBu^t)_8Cl(THF)_2$ are obtained.

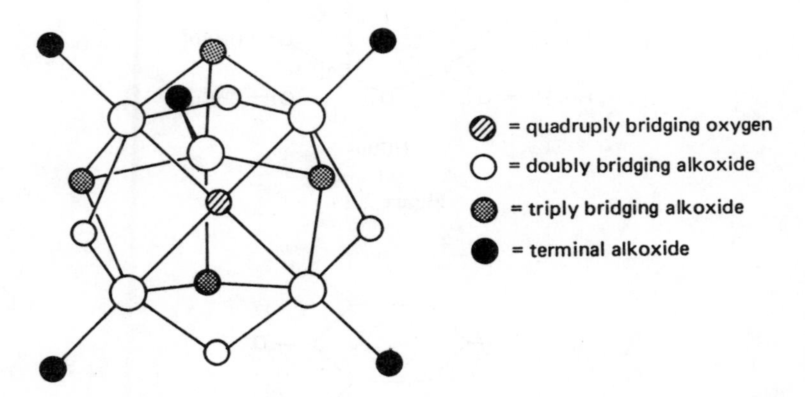

Figure 2.22 The coordination sphere in the alkoxides $M_5O(OPr^i)_3$ (M = Y, Yb); all metal atoms are 6-coordinate

Cluster formation is inhibited by the use of bulky ligands, forcing the metal to adopt a coordination number less than 6, thus:

$$Ce[N(SiMe_3)_2]_3 \xrightarrow[\text{pentane}]{Bu^t_3COH} Ce(OCBu^t_3)_3$$

This yellow solid is a monomer in solution and believed to contain 3-coordinate cerium; on thermolysis

$$2Ce(OCBu^t_3)_3 \xrightarrow{150^\circ} 2C_4H_8 + [Ce(OCHBu^t_2)_3]_2$$

it forms a dimeric alkoxide with four-coordinate Ce (Figure 2.23) and (largely) isobutene.

Silyloxides can also be made:

$$Y[N(SiMe_3)_2]_3 \xrightarrow[\text{toluene}]{Ph_3SiOH} Y(OSiPh_3)_3 \xrightarrow[(L\,=\,THF,py)]{L} Y(OSiPh_3)_3L_3$$

The adducts have *facial*-octahedral structures. (Similar compounds have been made for Ce, Pr, Nd.)

A range of 3-coordinate aryl oxides $M(OR')_3$ have been made using the bulky 2,6-di-tert-butyl-4-methylphenoxide ligand (R′) (Figure 2.24) (M = most lanthanides). Like the corresponding silylamides $Ln[N(SiMe_3)_2]_3$ (section 2.10.3) they probably have pyramidal rather than planar trigonal structures. They yield crystalline 1:1 adducts with Lewis bases (THF, Ph_3PO) which are doubtless 4-coordinate. A similar cerium compound $Ce(OR')_3$ (with X = H), which definitely has a pyramidal 3-coordinate structure, forms 1:1 adducts with THF, Et_3PO, $BuNH_2$ and Bu^tNC, as well as an interesting 5-coordinate bis-adduct

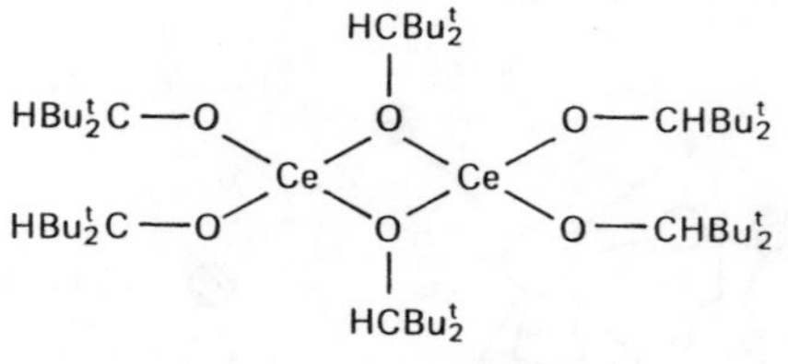

Figure 2.23

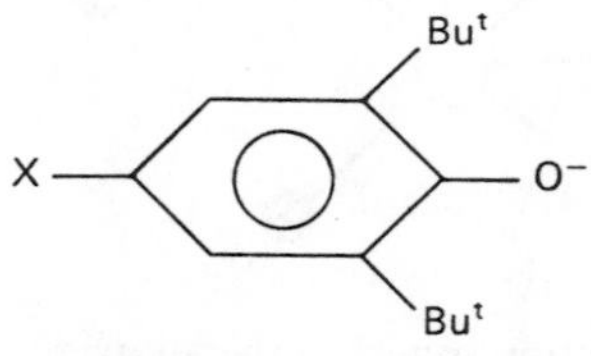

Figure 2.24

$Ce(OR')_3(Bu^tNC)_2$ (surprisingly, both isocyanides are on the same side of the O_3 plane). Using the sterically less demanding 2,6-dimethyl-phenoxide (OR″) affords two yttrium complexes:

$$YCl_3 + 3NaOR'' \xrightarrow{THF} Y(OR'')_3(THF)_3 \underset{THF}{\overset{toluene}{\rightleftharpoons}} [Y(OR'')_3(THF)]_2$$

The compound obtained depends on the recrystallising solvent. Their structures are shown in Figure 2.25.

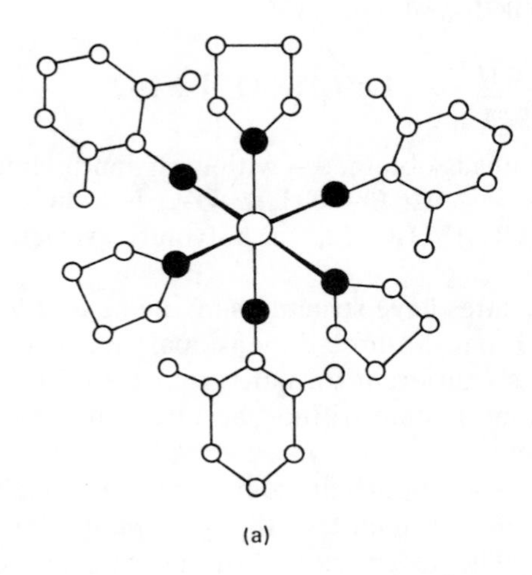

(a)

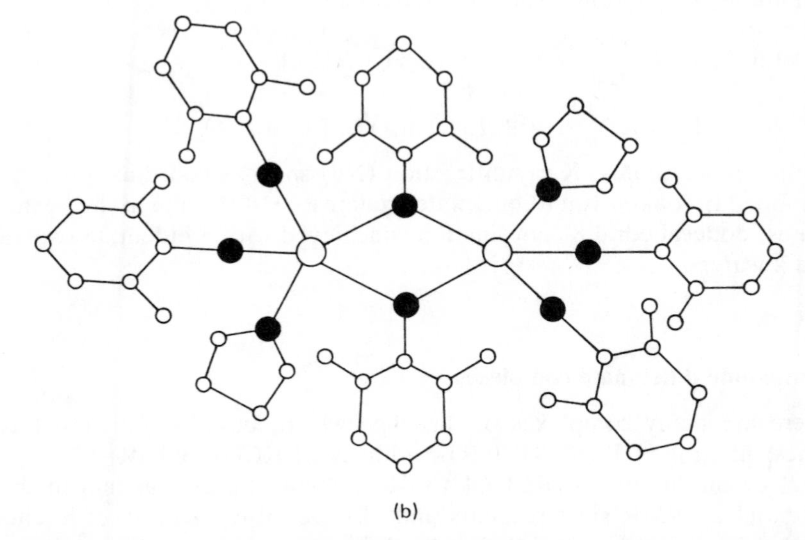

(b)

Figure 2.25 The 2,6-dimethylphenoxide complexes: (a) $Y(OR'')_3(THF)_3$; (b) $[Y(OR'')_3THF]_2$

The recent upsurge of interest in these compounds is due to their use as precursors for the deposition of pure metal oxides using either thermolytic or hydrolytic decomposition.

Complexes of carboxylic acids

Reaction of the lanthanide oxides with hot acetic acid affords hydrated acetates from which the water can usually be removed by azeotropic distillation with DMF and benzene:

$$Ln_2O_3 \xrightarrow[\text{heat}]{CH_3CO_2H} Ln(CH_3CO_2)_3.xH_2O$$

A variety of hydrates is obtained – with early lanthanides, x is usually 1.5, for most others $x = 4$. $Er(O_2CCH_3)_3.4H_2O$ is a dimer with 9-coordinate erbium; $Ce(O_2CCH_3)_3.H_2O$ has a polymeric structure, again with 9-coordinate cerium.

Other carboxylates have structures involving 8 or 9-coordination; thus $Y(HCOO)_3.2H_2O$ has a three-dimensional network with 8-coordinate yttrium involving six different formate groups. $Pr(CF_3COO)_3.3H_2O$ has 4 bridging and 1 monodentate trifluoroacetate and 3 coordinated waters to each praesodymium.

The oxalates are important compounds as they afford qualitative precipitation of the lanthanides, the precipitate being ignited to the weighable oxide. The exact nature of the precipitate depends on the conditions, but normal oxalates can be made:

$$Ln^{3+} \xrightarrow[\text{boil}]{H_2C_2O_4/Na_2C_2O_4} Ln_2(C_2O_4)_3.xH_2O$$

(Ln = La–Er, Y, $n = 9$; Er, Tm, Yb, Lu, $n = 6$)

In the nonahydrates, X-ray diffraction (Nd) shows 9-coordination with a tricapped trigonal prism (3 bidentate oxalates, $3H_2O$) while in the hexahydrates, dodecahedral 8-coordination is achieved with 3 bidentate oxalates and 2 waters.

Lanthanide diketonate complexes

There are many complexes of this type which, broadly, fall into three series: neutral $M(RCOCHCOR)_3$, adducts $M(RCOCHCOR)_3L_n$ ($n = 1$ or 2) or the anionic $M(RCOCHCOR)_4^-$. Their interest has lain in their potential as NMR shift reagents and also because volatility differences permit the separation of different lanthanides.

Typical preparations are:

$$NdCl_3 \xrightarrow{NH_4tfac(aq)} Nd(tfac)_3.2H_2O$$

$$Tb(NO_3)_3(aq) \xrightarrow{Hthd/NaOH} Tb(thd)_3$$

The compounds can be recrystallised from solvents like ethanol (or, in some cases, hydrocarbons); many can be sublimed *in vacuo*.

Table 2.8 Diketonate ligands $R_1COCHCOR_2$

R_1	R_2	Name	Abbreviation
CH_3	CH_3	acetylacetone	acac
CH_3	CF_3	trifluoracetylacetone	tfac
CF_3	CF_3	hexafluoroacetylacetone	hfac
Me_3C	CMe_3	dipivaloylmethane	dpm
		tetramethylheptanedione	tmhd
$CF_3CF_2CF_2$	CMe_3	1,1,1,2,2,3,3-heptafluoro-7,7-dimethyloctanedione	fod
2-thienyl (S)	CF_3	2-thenoyltrifluoroacetone	tta

Unlike the *d*-block metals, which form octahedral complexes $M(acac)_3$, the diketonates of the lanthanides are only rarely 6-coordinate. Thus $Ln(acac)_3(H_2O)_2$ (Ln = La, Pr, Nd, Eu, Ho, Y) have all been shown to be 8-coordinate (X-ray) while $Yb(acac)_3.H_2O$ is 7-coordinate. The water molecules are tightly bound; the complexes decompose on heating, but dehydration at room temperature yields Ln $acac_3$, apparently polymeric. Adducts with other Lewis bases can be isolated, such as the 7-coordinate Ln $acac_3.Ph_3PO$ and 8-coordinate Ln $acac_3$.phen.

Thenoyltrifluoroacetone complexes Ln tta_3 are involved in solvent extraction of lanthanides; like the lanthanides they form 8-coordinate Ln $tta_3.2H_2O$. A synergistic improvement in extraction is obtained in the presence of phosphine oxides and organophosphates, doubtless because of complexes like the 8-coordinate (X-ray) Nd $tta_3(Ph_3PO)_2$. Fluorinated chelates such as Nd $tfac_3$ and Ln $hfac_3$ are normally made as hydrates, but vacuum dehydration of the hexafluoroacetylacetonate affords the hygroscopic Ln $hfac_3$. These form thermally stable, volatile 8-coordinate adducts Ln $hfac_3(Bu_3PO)_2$ which can be separated by gas chromatography. The chelates Ln fod_3 are even more volatile; they associate in solution and readily hydrate to Ln $fod_3.H_2O$. Pr $fod_3.H_2O$ has an unusual 8-coordinate dimeric structure $[Pr_2fod_6(H_2O)].H_2O$ while Lu $fod_3.H_2O$ is 7-coordinate (Figure 2.26). They form many adducts Ln fod_3.L.

The ability to form adducts with a wide range of Lewis bases is necessary in the most important application of the fod chelates, as

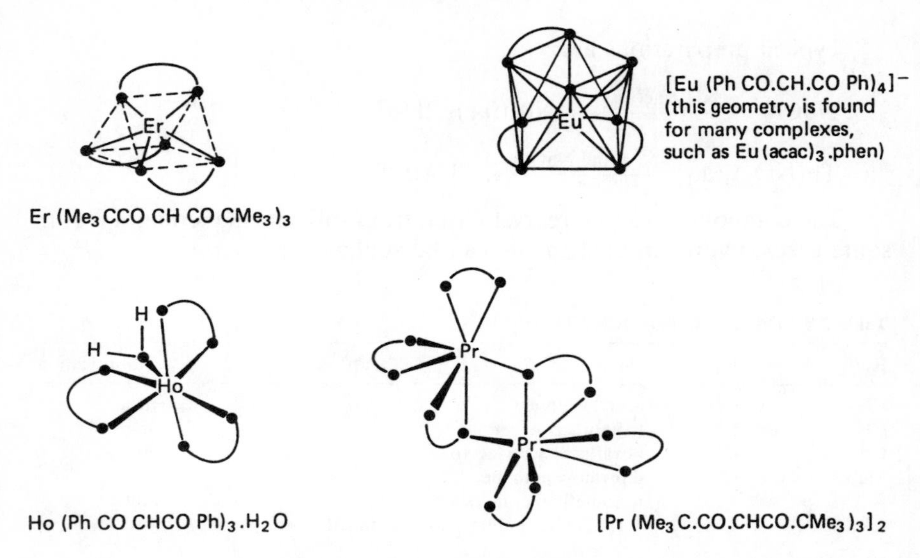

Figure 2.26 Structures of lanthanide complexes of β-diketonate ligands

lanthanide shift reagents, as for the hexane-soluble Ln dpm_3. The anhydrous chelates are dimers (7-coordinate) for the early lanthanides (Ln = La–Gd) and 6-coordinate monomers for the later members of the series. They are volatile *in vacuo* at 100–200°, the compounds of the heavier lanthanides being more volatile; presumably the smaller Ln^{3+} ions are better shielded, reducing intermolecular forces. The dpm chelates again form 7-coordinate hydrates Ln $dpm_3.H_2O$ (X-ray, Ln = Dy) and similar adducts M dpm_3.L for a range of other donors have been isolated (L = py, Ph_3PO, phen, bipy).

2.10.3 Complexes of nitrogen-donor ligands

The best characterised complexes are those with chelating ligands; the bidentate 2,2-bipyridyl, 1,10-phenanthroline and 1,8-naphthyridine, and the tridentate 2,2′,6,2″-terpyridyl. Even so, the complexes generally have to be prepared in a non-polar solvent such as ethanol.

Isolated complexes include $M(NO_3)_3L_2$ (L = bipy, phen; M = La–Lu), $MCl_3(bipy)_2(H_2O)_n$ (M = La–Nd, n = 1; M = Eu–Lu, n = 0), $La(phen)_4(ClO_4)_3$, M $phen_3(NCS)_3$ (M = La–Lu) and Yb-$phen_2(NCS)_3$; and $M(naphthy)_n(ClO_4)_3$ (M = La–Pr, n = 6; M = Nd–Eu, n = 5). Limited X-ray data show 10-coordination for $La(NO_3)_3$ $(bipy)_2$ and 12-coordination in $Pr(naphthyl)_6^{3+}$; discernible trends include the decrease in coordination number as the lanthanide ions become smaller, and higher coordination numbers with ligands like nitrate and 1,8-naphthyridine, which have a small ‘bite angle’. 9-coordination is found in the terpyridyl complexes $[M(terpy)_3](ClO_4)_3$ (M = La, Eu, Lu); where water and anio-

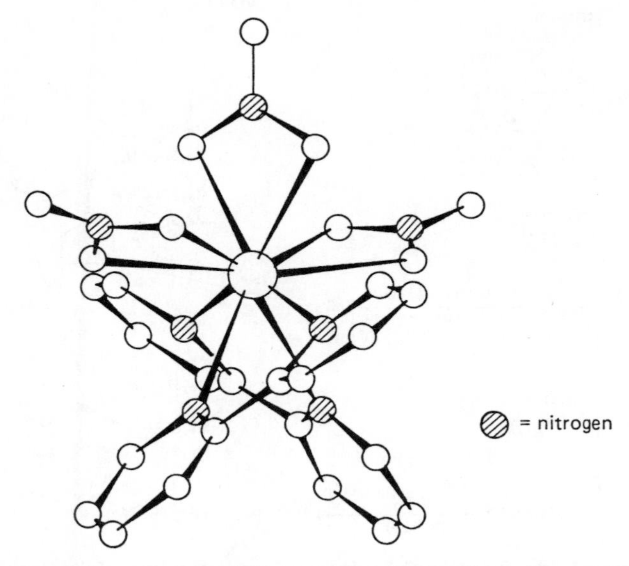

Figure 2.27 The structure of $La(bipy)_2(NO_3)_3$ [after A. R. Al-Karaghouli and J. S. Wood, *J. Amer. Chem. Soc.*, **90** (1968) 6548; reprinted with permission from the American Chemical Society © 1968]

nic ligands can both coordinate, a different form of 9-coordination is found in $[PrCl(terpy)(H_2O)_5]Cl_2.3H_2O$.

Alkylamide complexes

Reaction of anhydrous lanthanide chlorides with $LiN(SiMe_3)_2$ in THF affords the three-coordinate $Ln[N(SiMe_2)_2]_3$ (Ln = La–Lu, Y). Though moisture-sensitive, they are volatile (subliming *in vacuo* at 100°(10^{-4} mm) and pentane-soluble, with unusual pyramidal structures (like the scandium analogue). As well as synthetic utility for making alkoxides and alkyls, they form four-coordinate Ph_3PO adducts and most unusual peroxo-bridged species when excess phosphine oxide is used. The chlorine bridged $LuCl(N(SiMe_3)_2)_2$ compounds have been useful starting materials for making other derivatives, with mercaptans (Figure 2.28).

The corresponding di-isopropylamide ligand forms complexes $Ln[N(CHMe_2)_2]_3$, doubtless 3-coordinate but less sterically hindered, for unlike the corresponding silylamide, $Nd[N(CHMe_2)_2]_3$ forms a stable THF adduct.

Thiocyanate complexes

As expected for 'hard' acids, the lanthanides bind to the nitrogen of the NCS group. The stoichiometry of the anionic complexes depends on conditions:

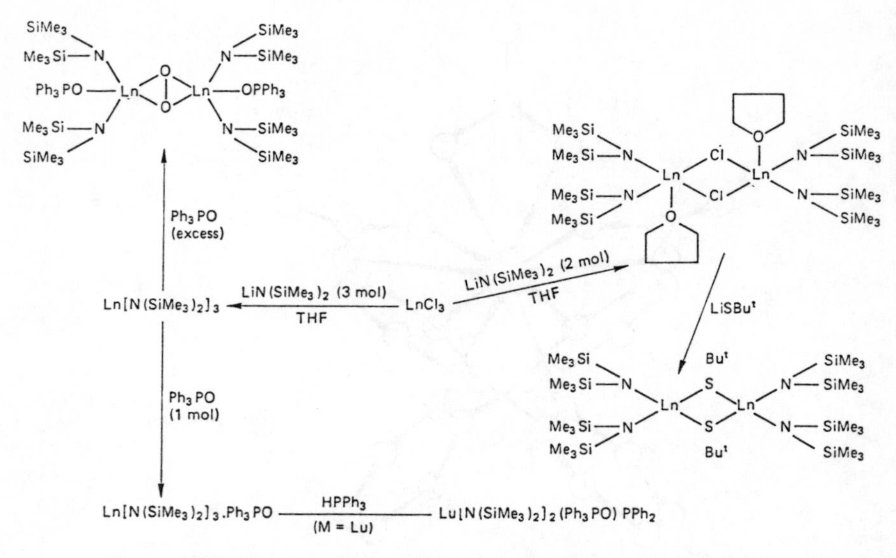

Figure 2.28 Some lanthanide bis(trimethylsilyl)amide complexes

$$Ln(NCS)_3 \xrightarrow[EtOH/Bu^tOH]{Bu_4NNCS} (Bu_4N)_3Ln(NCS)_6 \qquad (Pr–Yb, Y)$$

$$Ln(NCS)_3 \xrightarrow[MeOH/H_2O]{Et_4NNCS} (Et_4N)_4Ln(NCS)_7(H_2O) \qquad (La–Er)$$

$$Ln(NCS)_3 \xrightarrow[MeOH]{Me_4NNCS} (Me_4N)_4Ln(NCS)_7 \qquad (Dy, Er, Yb)$$

$(Bu_4N)_3Er(NCS)_6$ has octahedrally coordinated Er, the $Ln(NCS)_7(H_2O)^{4-}$ ion (Ln = Pr, Er) the rather rare cubic 8-coordination, and the $Ln(NCS)_7^{4-}$ ion is a distorted pentagonal bipyramid.

2.10.4 Complexes with halide ligands

Fluoride complexes are usually made by fusion methods. $NaLnF_4$ exists in two phases, the low-temperature phase exhibiting tricapped trigonal prismatic coordination (CN = 9) while at higher temperatures the fluorite structure (CN = 8) is adopted. With the larger alkali metals, the complexes M_3LnF_6 can also be made.

A series of hygroscopic complexes containing the LnX_6^{3-} ion (X = Cl, Br, I) have been made:

$$ErCl_3 \xrightarrow[ErOH]{Ph_3PHCl/HCl} (Ph_3PH)_3ErCl_6$$

$$NdBr_3 \xrightarrow[HBr]{pyHBr/EtOH} (PyH)_3NdBr_6$$

$$(Ph_3PH)_3SmBr_6 \xrightarrow[(anhyd)]{HI} (Ph_3PH)_3SmI_6$$

The iodides are particularly unstable. These complexes are a rare example of octahedral coordination and are notable for weak transitions in the electronic spectra. Other complexes, like $Cs_2NaLnCl_6$ (elpasolite structure) have been made by fusion; a range of stoichiometries is obtained. A coordination number of 6 is the rule, as in Cs_2DyCl_5 (octahedra share *cis*-corners) and $Cs_3Y_2I_9$, like $Cs_3Ln_2Br_9$.

2.10.5 Complexes with sulphur-containing ligands

These are predominantly obtained with chelating ligands. $M(S_2P(C_6H_{11})_2)_3$ (M = Pr, Sm) are 6-coordinate (trigonal prismatic) but most are 8-coordinate, with a range of geometries – $Er(S_2P(OEt)_2)_4^-$ and $Eu(S_2CNEt_2)_4^-$ (dodecahedral), $Pr(S_2PMe_2)_4^-$ (tetragonal antiprism), and $La(S_2P(OEt)_2)_3(Ph_3PO)_2$ (distorted square antiprism). With neodymium and smaller lanthanides, complexes like the last named are not obtained, instead the 7-coordinate $(Ln\{S_2(OEt)_2\}_2(OPPh_3)_3]^+$ (M = Nd–Lu) ions are obtained.

2.10.6 Complexes of EDTA and related ligands

Complexes of the type M^+ $[Ln(EDTA)\ (H_2O)_x]^-.nH_2O$ can readily be prepared thus:

$$Ln_2O_3 + H_4EDTA + NaOH \xrightarrow{100^\circ} Na^+[Ln(EDTA)(H_2O)_x]^-.nH_2O$$

They demonstrate the lanthanide contraction as the 9-coordinate ion $[Ln(EDTA)(H_2O)_3]^-$ is formed for La–Dy, while $[Ln(EDTA)(H_2O)_2]^-$ (8-coordinate) is found for Er and Yb (Figure 2.29). The diethylenetriaminepentaacetate ligand can be octadentate, and in the complex ions $[Ln(DTPA)(H_2O)]^-$ (Ln = Gd etc.), another 9-coordinate (capped square antiprism) species is found.

There is considerable current interest in the use of complexes like $[Gd(DTPA)(H_2O)]^-$ as contrast agents for diagnostic imaging in NMR body scanners, to assist *in vivo* tissue characterisation. The proton NMR signals come largely from water molecules; strongly paramagnetic ions shorten relaxation times and enhance signals. Use of the chelating ligand DTPA leads to the formation of a very stable ($\log K \sim 22$) complex with low toxicity, excreted intact by the kidneys (the EDTA complex, log $K \sim 17$, is quite toxic). Clinical trials of this complex have been particularly concerned with detecting brain tumours. Other complexes, including those of neutral ligands, are currently under investigation.

Crown ether-type ligands containing sulphur or nitrogen as the donor atoms also form isolable complexes.

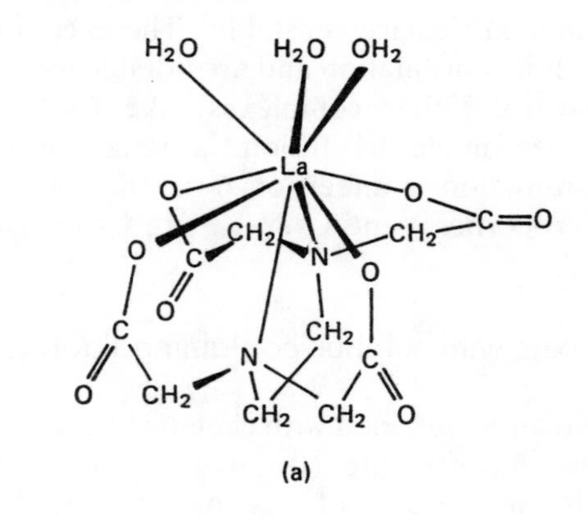

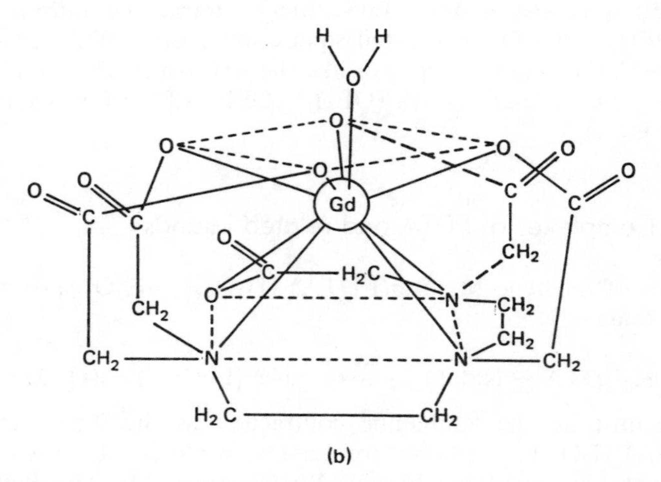

Figure 2.29 (a) The 9-coordinate $La(EDTA)(OH_2)_3^-$ ion. (b) The 9-coordinate $GdDTPA(OH_2)^-$ ion with distorted capped square antiprismatic coordination [adapted from H. Gries and H. Miklautz, *Physiol. Chem. Phys. Med. NMR*, **16** (1984) 105]

2.10.7 Complexes of macrocyclic ligands

(a) Crown ethers

A considerable number of these have been synthesised in recent years from the reaction of the crown ether with a lanthanide salt in a non-aqueous solvent like MeCN. In most cases, the lanthanide is coordinated to the donor atoms (usually oxygen) in the crown 'ring' although compounds are known where a complex ion is hydrogen-bonded to the crown ether so that there are no direct bonds between Ln and the crown ether.

Coordination numbers are generally high in these complexes, not least because many are lanthanide nitrate complexes, nitrate invariably being bidentate. Typical examples are: M(18-crown-6)$(NO_3)_3$ (M = La, Nd; CN = 12); M(12-crown-4)$(NO_3)_3$ (M = Nd–Lu, CN = 10); Eu(15-crown-5)$(NO_3)_3$ (CN = 11). The complex $ErCl_3.5H_2O$.12-crown-4 is in fact [Er(12-crown-4)$(H_2O)_5]^3(Cl^-)_3$ (CN = 9). However $YCl_3.8H_2O$.15-crown-5 is $[Y(OH_2)_8]^{3+}(Cl^-)_3$.15-crown-5 with no direct Y–crown interaction, a consequence of the small yttrium ion. Similar steric factors mean that only the lighter lanthanides form stable complexes with the larger 18-crown-6 ligand.

Sometimes, depending on the reaction stoichiometry, more than one complex can be isolated. Thus, apart from its 1:1 complex with 18-crown-6, neodymium nitrate also forms $Nd(NO_3)_3$.(18-crown-6)$_{0.75}$, which is $[Nd(NO_3)_2$(18-crown-6)$]_3Nd(NO_3)_6$ with 10 and 12-coordinate Nd in the cation and anion respectively. Figure 2.30 shows typical ligands.

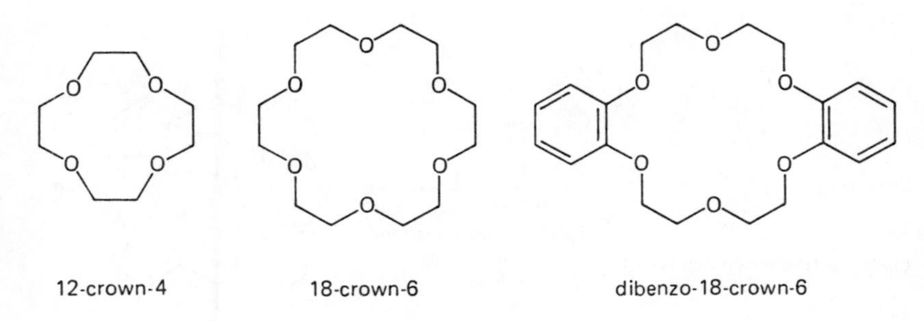

Figure 2.30

(b) Cryptands

These differ from crown ethers in that they encapsulate the metal ion. A typical example of these ligands is 2,2,2-cryptate, shown in Figure 2.31. The ligand is potentially octadentate, thus in the complex $[La(NO_3)_2$(2,2,2-cryptate)$]_3[La(NO_3)_6]$ the lanthanums in both the anion and cation are 12-coordinate.

(c) Porphyrins and phthalocyanines

These may be prepared by typical routes (Figure 2.32).

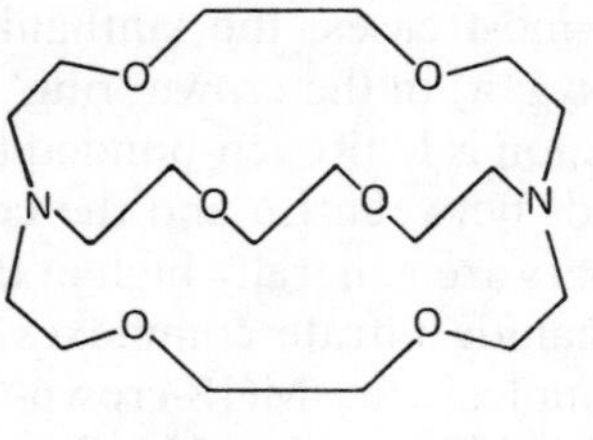

Figure 2.31

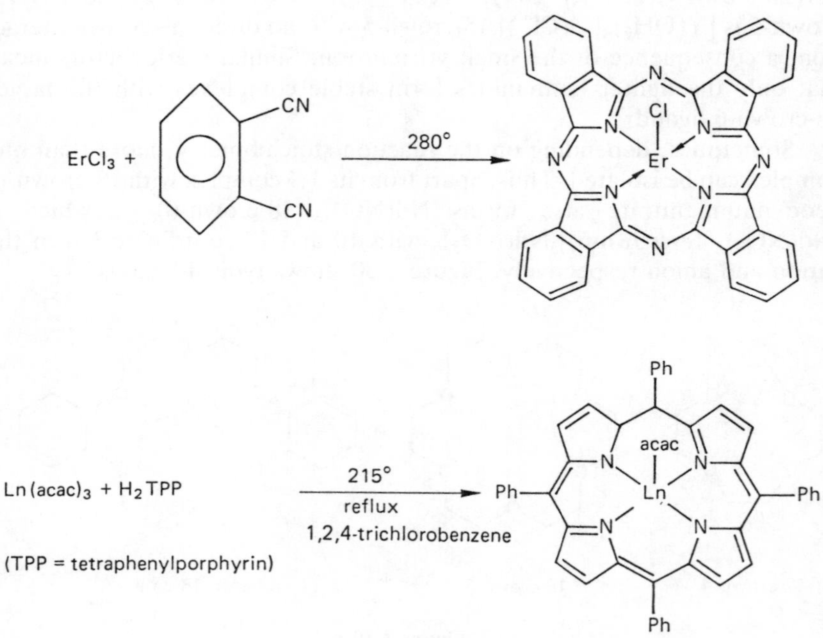

$ErCl_3$ + [phthalonitrile] $\xrightarrow{280°}$ [Er phthalocyanine chloride]

$Ln(acac)_3 + H_2TPP$ $\xrightarrow[\text{reflux, 1,2,4-trichlorobenzene}]{215°}$ [Ln(acac)TPP]

(TPP = tetraphenylporphyrin)

Figure 2.32

2.11 Some applications of lanthanides in NMR

There has been increasing interest over the last 25 years in utilising paramagnetic lanthanide ions in the rationalisation of NMR spectra of organic compounds and in the elucidation of their structures.

2.11.1 Lanthanide Shift Reagents (LSRs)

These are lanthanide compounds which, when added to a solution of an organic compound (such as alcohols, amines) cause a spreading out (shift) of the proton resonances. LSRs can be utilised to shift the resonances of other nuclei too. Most reports have centred upon 6-coordinate β-diketonate complexes of bulky ligands, such as $Ln(Me_3C.COCHCOCMe_3)_3$ (Ln = Eu, Pr, for example).

There are several important factors that contribute to these compounds acting as shift reagents:

(a) they are paramagnetic
(b) they act as Lewis acids
(c) they have vacant coordination sites for Lewis bases
(d) lanthanides commonly adopt coordination numbers greater than 6.
(e) they are soluble in covalent NMR solvents like CCl_4.

Thus the addition of a small amount of one of the complexes mentioned above to a solution of, say, pentanol can simplify the ^1H NMR spectrum into a first-order spectrum with virtually no effect on the coupling constants (Figure 2.33). The europium compound shifts the resonances to low field while the praseodymium compound shifts them in the opposite sense. The effect is greatest on resonances nearest the donor atom in the organic molecule.

The shifts (often called LIS, Lanthanide Induced Shifts) produced by LSRs are primarily pseudo-contact (through space) rather than contact (delocalisation of unpaired spin density from the metal on to the organic molecule) but it is necessary to separate the contributions in each case. (Ideally this can be done by finding the shift produced by the gadolinium compound, which is entirely contact; in fact, the line-broadening usually produced can prevent resonances from being observed.)

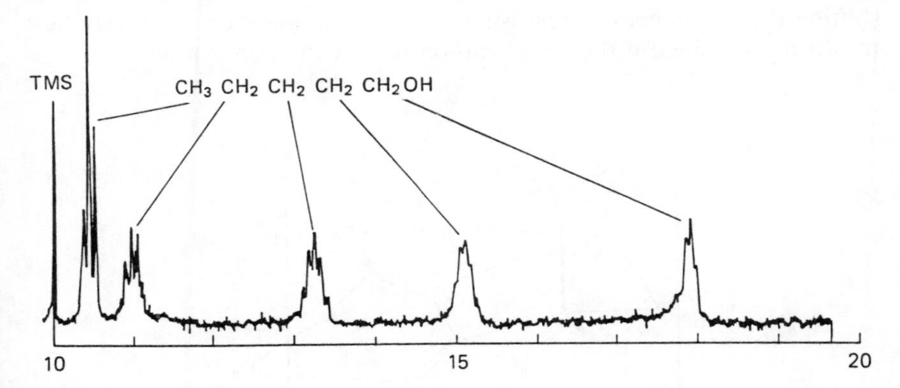

Figure 2.33 100 MHz ^1H NMR spectrum of n-pentanol (0.22 M) and $Pr(tmhd)_3$ (0.053 M); CCl_4 solution, τ units, internal Me_4Si standard [from J. Briggs *et al.*, *Chem. Comm.*, (1970)749; reprinted by permission of the Royal Society of Chemistry]

The pseudo-contact shift LIS_{pc} (= $\Delta H/H$) for any one nucleus can be related to the molecule geometry via the equation

$$LIS_{pc} = \frac{K(3\cos^2\theta - 1)}{r^3}$$

where K is a constant, θ is the angle relating the proton position to the symmetry axis of the complex (polar coordinate), and r is the proton–lanthanide distance.

The equation assumes axial symmetry for the LSR-adduct complex; this is not usually justified on 'static' crystallographic grounds but is generally valid in solution owing to fluxionality – rapid rotation about the Ln–O bond in solution. It also assumes that only one adduct is present; though it does appear that 1:1 adducts predominate in solution, especially with bulky Lewis bases, both 1:1 and 1:2 adducts have been isolated in the solid state.

Some molecules which do not form adducts with the tris(β-diketonates) can be studied with the help of the binuclear silver(I)–lanthanide(III) reagents like $Ag^+Ln(fod)_4^-$ (Ln = Pr, Yb; fod = $CF_3CF_2CF_2COCH.COCMe_3$) in which the silver binds to the 'soft' Lewis base adduct, and then forms an ion-pair with Ln fod_3 so that the resonances in the organic molecule are shifted. This has been used to analyse mixtures of xylenes or other arenes, and mixtures of *cis*- and *trans*-alkenes.

Another variant is the chiral shift reagent like the substituted camphor derivative $Eu(facam)_3$ (Figure 2.34) which will bind to one enantiomer in a different way – either more strongly or in a different geometry – thus shifting the resonances of one isomer more than another; obviously best results are obtained if the chiral carbon is near the donor atom.

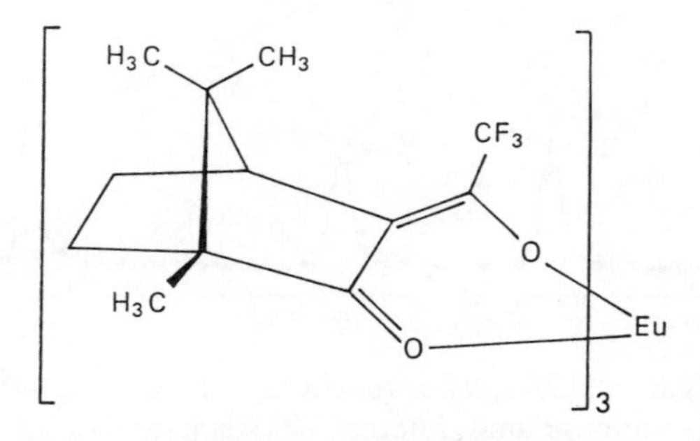

Figure 2.34 The chiral shift reagent $Eu(facam)_3$

Shift reagents in aqueous solution

The insolubility of the β-diketonate complexes in water has meant the use of 'ionic' lanthanide salts (like the chloride and nitrate) as shift reagents. The simple aquo-complexes do however tend to precipitate in neutral or alkaline solution so that for use at pH values greater than 5, complexes like $[Ln(EDTA)(H_2O)_3]^-$ – stable at pH 6–7 but still having vacant sites – have been used in the study of molecules like adenosine – and cytosine – 5′-monophosphate. As anionic complexes obviously tend to associate with cations, paramagnetic complexes like $Ln(hpda)_3^{6-}$ (hpda = 4-hydroxypicolinic acid) or $Ln(dota)^-$ (Figure 2.35) produce shifts in the NMR spectra of ions like $^{23}Na^+$, $^{24}Mg^{2+}$, $^{39}K^+$, $^{87}Rb^+$, $^{14}NH_4^+$ and $^2H^+$. Since cells and tissues contain high concentrations of Na^+ and K^+, this has applications to *in vivo* studies – the complex $Dy(DOTP)^{5-}$ (Figure 2.36) has been used to study $^{23}Na^+$ in living rats' hearts.

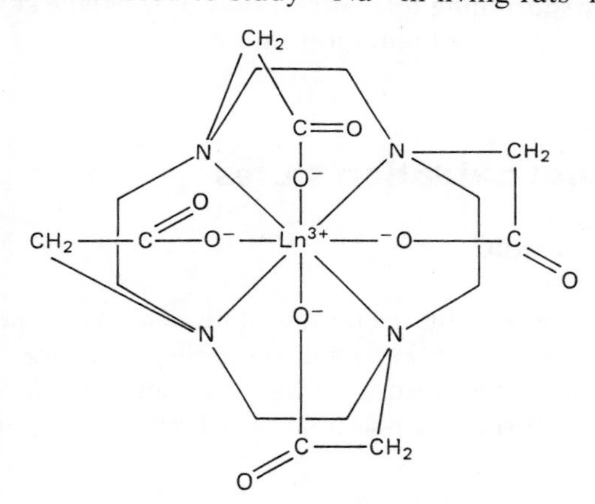

Figure 2.35 $Ln(dota)^-$

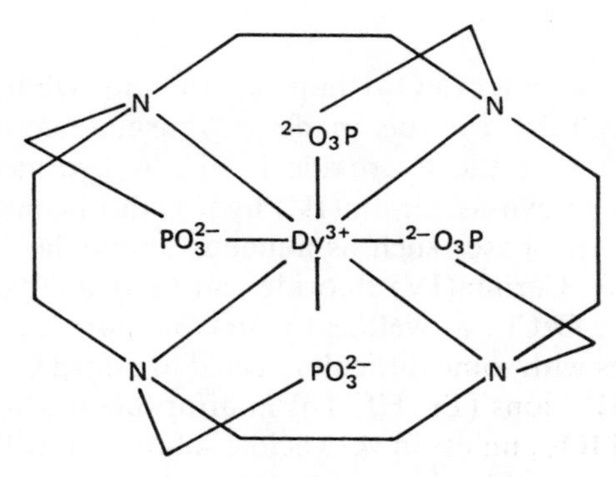

Figure 2.36 $Dy(DOTP)^{5-}$

The ability of lanthanides to substitute for ions like Ca^{2+} means that they have potential in the study of metalloproteins, but difficulties that have to be overcome include the problems of overlapping resonances and poor resolution, as well as the possibility of more than one binding site.

Magnetic resonance imaging

The use of gadolinium complexes in this area is also referred to in section 2.10.6. Magnetic resonance imagers are basically pulsed FT NMR spectrometers; they use gradient coils to create different fields at many points in a given piece of tissue, thus yielding resonances at slightly different frequencies at each point, with a computer processing the data to give a digitised image. At the present time, spectra are obtained for protons in water molecules, and the role of the lanthanide is to give signal intensity enhancement by shortening relaxation times. For *in vivo* study, complexed Gd^{3+} is used, as free gadolinium ions are toxic.

2.12 Unusual oxidation states

2.12.1 The +4 state

Five lanthanides adopt this oxidation state. Of these, neodymium and dysprosium only form fluoride complexes in the solid state; praseodymium and terbium additionally form a tetrafluoride and dioxide. Cerium is the only lanthanide to form water-soluble coordination compounds in the +4 state.

2.12.2 Cerium (+4)

Of the binary compounds, CeO_2 (fluorite structure; white when stoichiometric, usually yellow) can be made by burning cerium or igniting cerium(III) oxy-salts or the hydroxide in air. A hydrated form, yellow $CeO_2.xH_2O$ (also known as cerium(IV) hydroxide) is made by precipitation of Ce^{4+}(aq) with bases such as aqueous ammonia. CeO_2 is used in self-cleaning ovens. Cerium(IV) fluoride can be made by fluorination of the metal, CeF_3 or $CeCl_3$, as well as by precipitation.

CeO_2 dissolves with some difficulty in acid to afford Ce^{4+}(aq). This ion resembles other M^{4+} ions (Zr, Hf, Th) in affording precipitates with ions like F^-, PO_4^{3-} and IO_3^-; unless in very acidic solution, it will be hydrolysed, and in most acids a complex-ion rather than the aquo-ion will be present, accounting for the variation in $E^0(Ce^{4+}/Ce^{3+})$: 1.44 V (1M H_2SO_4), 1.6 V (1M HNO_3), 1.7 V (1M $HClO_4$).

In view of the potential for the oxidation of water:

$$2H_2O \longrightarrow O_2 + 4H^+ + 4e^- \qquad E^0 = -1.23\ V$$

the stability of Ce^{4+} in solution must be due to kinetic rather than thermodynamic factors.

Ceric sulphate and ceric nitrate can be produced as crystalline salts, $Ce(SO_4)_2.4H_2O$ and $Ce(NO_3)_4.5H_2O$, by dissolving ceric oxide in a concentrated solution of the appropriate acid. The nitrate is isomorphous with the thorium analogue, thus presumably containing 11-coordinate cerium. It is more normally met with as the orange complex ion $Ce(NO_3)_6^{2-}$ (bidentate nitrates, 12- coordinate cerium):

$$CeO_2.xH_2O \xrightarrow{8\text{–}16M\ HNO_3} Ce(NO_3)_4(aq) \xrightarrow{NH_4NO_3} (NH_4)_2Ce(NO_3)_6$$

This has long been used as a standard oxidant in titrimetry and as an oxidising agent in organic chemistry. Other anionic complexes include $(NH_4)_4Ce(SO_4)_2.2H_2O$ and $Na_6Ce(CO_3)_5.12H_2O$ (10-coordinate).

Several fluoride complexes can be synthesised thus:

$$CeF_4 \xrightarrow[H_2O]{exc\ NH_4F} (NH_4)_4CeF_8 \xrightarrow{heat} (NH_4)_2CeF_6$$

The CeF_8^{4-} ion is a distorted square antiprism, a similar geometry also being found in the chain structure of $(CeF_6^{2-})_x$. $(NH_4)_3CeF_7.H_2O$ contains $Ce_2F_{14}^{6-}$ ions (8-coordinate, dodecahedral). The water-sensitive alkali metal fluorides are made by dry methods:

$$MCl + CeO_2 \xrightarrow{F_2} M_2CeF_6,\ M_3CeF_7\ (M = Na, K, Rb, Cs)$$

$CeCl_4$ does not exist, owing to the oxidising power of Ce^{4+}, but salts of the yellow $CeCl_6^{2-}$ can be made:

$$CeO_2.xH_2O \xrightarrow[diglyme]{SOCl_2} H_2diglyme_3\ CeCl_6$$

$$CeO_2.xH_2O \xrightarrow[EtOH/RCl]{HCl(g)} R_2CeCl_6\ (R = Ph_3PH, Et_4N, Ph_4As)$$

These can be converted to the violet, moisture-sensitive $CeBr_6^{2-}$:

$$CeCl_6^{2-} \xrightarrow{HBr(g)} CeBr_6^{2-}$$

Reduction occurs on heating:

$$2Cs_2CeCl_6 \xrightarrow{300^\circ} 4CsCl + 2CeCl_3 + Cl_2$$

Various O-donor ligands afford orange to yellow complexes:

$$(NH_4)_2Ce(NO_3)_6 \xrightarrow[Me_2CO]{Ph_3PO} Ce(NO_3)_4(Ph_3PO)_2 \qquad \text{(10-coord.)}$$

$$(NH_4)_2Ce(NO_3)_6 \xrightarrow[Ph_3PO]{HCl(g)/MeOH} CeCl_4(Ph_3PO)_2 \quad (\textit{cis})$$

$$H_2(diglyme)_3CeCl_6 \xrightarrow[EtOAc(MeOH)]{L} CeCl_4L_2 \quad (\textit{trans})$$

$$(L = Ph_3AsO, Bu^t_2SO, (Me_2N)_2CO, (Me_2N)_3PO)$$

$CeCl_4(Me_2SO)_3$ can also be made by the latter method and is probably $CeCl_2(Me_2SO)_6^{2+}CeCl_6^{2-}$.

A number of cerium(IV) β-diketonates, with ligands like acetylacetone and dibenzoylmethane, have been made by oxidation of the Ce(III) analogues:

$$Ce(NO_3)_3(aq) + 3R_1.COCH.COR_2 + 3NH_3$$
$$\longrightarrow Ce(R_1COCHCOR_2)_3 \xrightarrow{O_2} Ce(R_1COCHCOR_2)_4$$

They are dark-red crystalline solids (8-coordinate, square antiprismatic), soluble in solvents like benzene and chloroform, and have attracted attention as lead-free antiknock agents.

Another class of essentially covalent Ce(IV) compounds is provided by the alkoxides $Ce(OR)_4$:

$$(pyH)_2CeCl_6 + 4ROH + 6NH_3 \longrightarrow Ce(OR)_4 + 2py + 6NH_4Cl$$

$$(NH_4)_2Ce(NO_3)_6 + 4ROH + 6NaOMe$$
$$\longrightarrow Ce(OR)_4 + 6NaNO_3 + 2NH_3 + 6MeOH$$

A coordination number of 4 is too low for monomers, so that many of the compounds are involatile and polymeric. $Ce(OPr^i)_4$ is volatile *in vacuo* below 200°. Monomers include $Ce(OR)_4(THF)_2$ ($R = CMe_3, SiPh_3$) while 6-coordination is also found in $Ce(OCMe_3)_6^{2-}$.

2.12.3 Other metals

In general, the higher oxides obtained for praseodymium and terbium are usually represented as Pr_6O_{11} and Tb_4O_7, but the phase made depends on conditions:

$$Tb(NO_3)_3 \xrightarrow{heat/O_2} Tb_4O_7; \qquad Tb_2O_3 \xrightarrow[450^\circ]{O} TbO_2$$

$$Pr(NO_3)_3 \xrightarrow{heat} Pr_2O_3 \xrightarrow[O_2]{heat} Pr_6O_{11} \xrightarrow[450^\circ]{O} PrO_2$$

Both are soluble in acids, but because of the higher M^{3+}/M^{4+} potentials, they oxidise the acids and solutions of Ln^{3+} are formed.

The fluorides are made thus:

$$TbF_3 \xrightarrow[300°]{F_2} TbF_4; \quad Na_2PrF_6 \xrightarrow{\text{dry HF}} PrF_4$$

As with cerium, fluorocomplexes can be made by dry methods:

$$Cs_3LnCl_6 \xrightarrow{XeF_2} Cs_3LnF_7 \text{ (Ln = Pr, Tb, Dy, Nd)}$$

$$MCl + Ln_xO_y \xrightarrow{F_2} M_3LnF_7, M_2LnF_6 \text{ (M = Na–Cs; Ln = Tb, Pr)}$$

2.12.4 The +2 state

Europium, and, to a lesser extent, samarium and ytterbium, form reasonably stable M^{2+}(aq) ions, though they are oxidised by atmospheric oxygen. They are obtained either by electrolytic reduction or chemical reduction with alkali metal amalgams, though Eu^{2+} is produced by milder reducing agents (such as Zn). Several other metals (such as Dy, Tm, Nd) form stable dihalides (Section 2.8.1) but these are immediately oxidised by water. The stability of divalent ions is enhanced by the use of non-aqueous solvents such as THF or HMPA (typical colours are: Eu^{2+} pale yellow; Sm^{2+} red; Yb^{2+} green-yellow; Tm^{2+} green; Dy^{2+} brown; Nd^{2+} red).

The M^{2+}(aq) ions are precipitated by anions such as CO_3^{2-}, SO_4^{2-}, CrO_4^{2-}, $C_2O_4^{2-}$ or OH^-, showing a general resemblance to the alkaline earth metals (the ionic radius of Sr^{2+} is similar to that of Eu^{2+} and Sm^{2+}); the carbonates have the aragonite structure, the sulphates generally the barium sulphate structure, while the oxalate $EuC_2O_4.H_2O$ is isostructural with the strontium analogue. These compounds are stable in dry air. Traditionally, europium can be separated from the other lanthanides by zinc (powder) reduction to Eu^{2+} followed by coprecipitation with barium on the addition of $BaCl_2$ and H_2SO_4.

Several kinds of binary compound have been mentioned in section 2.8; many chalcogenides are of interest for their magnetic properties. The compounds MX generally have the NaCl structure, in general those of Sm, Yb and Eu being genuine divalent compounds while others are semimetallic, $M^{3+}(X^-)_2(e^-)$.

Limited study has been made of simple Lewis base adducts, such as $MCl_2(THF)_x$ (M = Nd, x = 2; M = Yb, x = 1), $SmI_2(Bu^tCN)_2$ (6-coordinate polymer) and $SmI_2[O(C_2H_4OMe)_2]_2$ (8-coordinate, existing as *cis*- and *trans*-isomers).

Among the best-characterised divalent compounds are those with bulky alkylamide and alkoxide ligands (Figure 2.37) where coordination numbers of 3 and 4 predominate. The trimethylalumium adduct of (bistrimethylsilylamido)ytterbium is not 2-coordinate in reality, as short Yb–C contacts supplement the Yb–N bonds. While $M(OR)_2L_2$ (L = THF, Et_2O) have tetrahedral structures, $M(OR)_2(THF)_3$ is, unexpectedly, square-pyramidal.

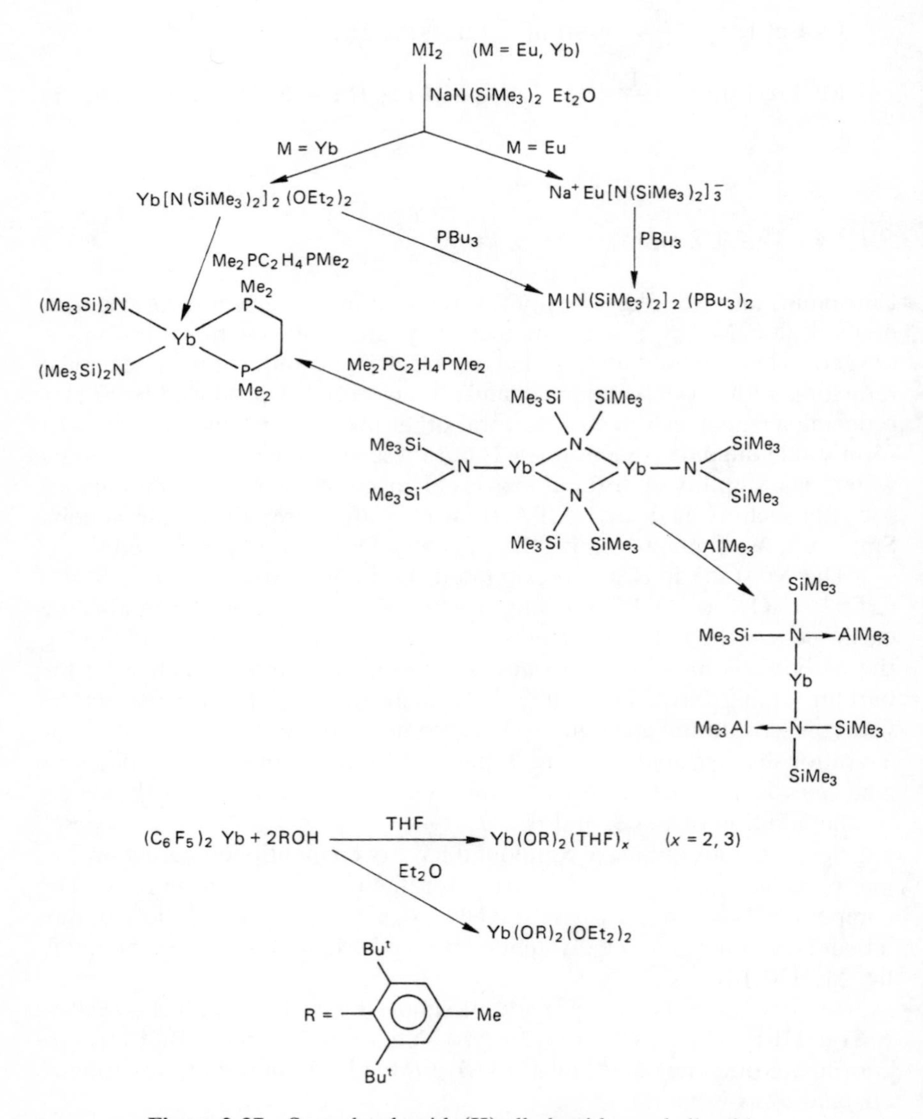

Figure 2.37 Some lanthanide(II) alkylamides and alkoxides

2.13 Promethium

In 1914, Moseley predicted the existence of an element between neodymium and samarium. Attempts to identify it in lanthanide ores were fruitless, as all 13 promethium isotopes are radioactive (with short half-lives). Early synthetic attempts involving neutron-irradiation of neodymium had to await the advent of ion-exchange chromatography before pure promethium could be separated (Marinsky and Coryell, 1945).

$$^{146}_{60}Nd \xrightarrow{n,\ \gamma} {}^{147}_{60}Nd \xrightarrow[11\alpha]{\beta^-} {}^{147}_{61}Pm$$

Promethium is also formed by spontaneous fission of ^{238}U and thus obtained by work-up of the fission products from nuclear reactors. Hence it is present at low levels in uranium ores – a sample of African pitchblende was found to contain 4×10^{-15} g of ^{147}Pm per kg of ore. ^{147}Pm (β^-, 2.64 y) is the most useful isotope; ^{145}Pm (17.7 y) is the longest lived.

Chemistry of the element

A number of promethium compounds, usually red or pink, have been synthesised using standard methods:

$$Pm^{3+}(aq) \xrightarrow{HF} PmF_3.xH_2O \xrightarrow[300°]{F_2} PmF_3 \quad \text{purple-pink (mp 1338°)}$$

$$Pm_2O_3 \xrightarrow[500°]{HCl(g)} PmCl_3 \quad \text{lavender (mp 655°)}$$

$$Pm_2O_3 \xrightarrow[500°]{HBr(g)} PmBr_3 \quad \text{coral-red (mp 625°)}$$

$$PmX_3 \xrightarrow[400°]{HI(g)} PmI_3 \quad \text{red (mp 695°)}$$

(X = Cl, Br)

The trihalides have, successively, the LaF_3 (X = F), UCl_3 (X = Cl) and $PuBr_3$ (X = Br, I) structures. The iodide is dimorphic, having the 6-coordinate BiI_3 structure at high temperatures. They are soluble, except for the fluoride. On the other hand, both the oxalate and hydroxide are insoluble:

$$PmCl_3(aq) \xrightarrow{H_2C_2O_4(aq)} Pm_2(C_2O_4).10H_2O \text{ (rose-violet)}$$

$$Pm^{3+}(aq) \xrightarrow{NH_3(aq)} Pm(OH)_3 \quad \text{(pink)}$$

Both afford the oxide on heating, for example

$$Pm_2(C_2O_4)_3.10H_2O(s) \xrightarrow{600^\circ} Pm_2O_3(s) + 3CO(g) + 3CO_2(g) + 10H_2O(g)$$

Some other salts have been made but not fully characterised, such as a nitrate and hydrated chloride, probably $PuX_3.6H_2O$ (X = NO_3, Cl).

Metallic promethium has been obtained as a silvery solid (mp 1080°) by metallothermic reduction:

$$PmF_3 \xrightarrow[1000^\circ]{Li} Pm$$

Thus, although relatively little is still known about it, the chemistry of promethium is a perfect microcosm of lanthanide chemistry.

2.14 Organometallic chemistry

The greatest developments in lanthanide chemistry in the past twenty years have taken place here. The principal differences between the lanthanides and the *d*-block metals lie in the lack of stable carbonyls and in the reactivity of the compounds, not infrequently pyrophoric in air.

2.14.1 σ-bonded compounds

There are several kinds of these, all having coordinative saturation in common. This may be achieved (i) using bulky ligands, sometimes via the formation of anionic derivatives, (ii) with chelating ligands, or (iii) using neutral ligands like THF to 'block' vacant sites. Thus in class (i) we have the three-coordinate $M[CH(SiMe_3)_2]_3$ (M = La, Sm; pyramidal like the isoelectronic silylamides discussed in section 2.8.3) and $LiL_4^+MR_4^-$ (L = THF or Et_2O; R = CMe_3, $CH(SiMe_3)_2$ or 2,6-dimethylphenyl).

In general, they are prepared by salt-elimination reactions:

$$MCl_3 + 6LiCH_3 \xrightarrow[Et_2O]{Me_2N.C_2H_4.NMe_2} [Li(Me_2NC_2H_4NMe_2)^+]_3M(CH_3)_6^{3-}$$

(M = La–Lu, Y, except Eu)

$$MCl_3 + 4Li\text{-}C_6H_3(CH_3)_2 \xrightarrow{THF} Li(THF)_4^+M(C_6H_3(CH_3)_2)_4^-$$

(M = Yb, Lu)

The $Er(CH_3)_6^{3-}$ ion is octahedral (X-ray) while the tetrahedral arylate ion is shown in Figure 2.38. Ylids and other chelating ligands afford 6-coordinate complexes like those shown in Figure 2.39.

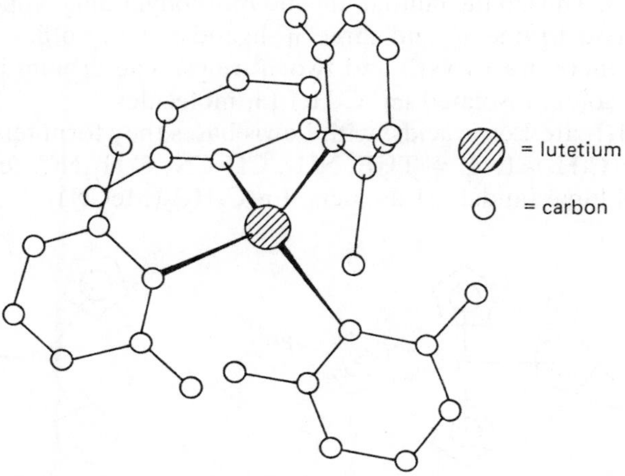

Figure 2.38 The Lu(2,6-dimethylphenyl)$_4^-$ ion, showing how the use of bulky ligands leads to the unusually low coordination number of 4 for lutetium [after S. A. Cotton *et al.*, *Chem. Comm.*, (1972) 1225; reprinted with permission of the Royal Society of Chemistry]

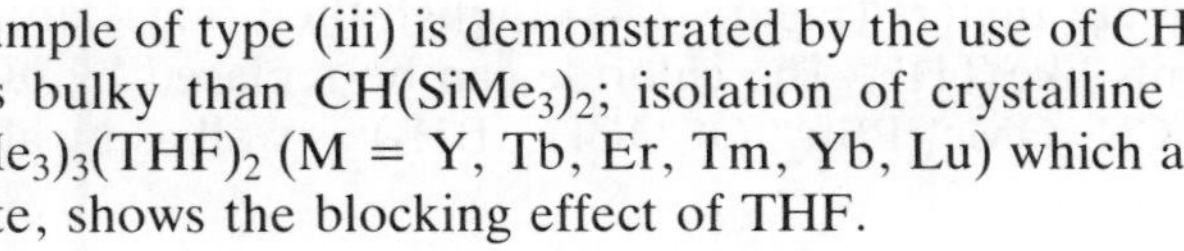

R = Me, Bu^t for example

Figure 2.39

An example of type (iii) is demonstrated by the use of CH_2SiMe_3 as a ligand, less bulky than $CH(SiMe_3)_2$; isolation of crystalline compounds $M(CH_2SiMe_3)_3(THF)_2$ (M = Y, Tb, Er, Tm, Yb, Lu) which are probably 5-coordinate, shows the blocking effect of THF.

2.14.2 Cyclopentadienyl complexes

These were the first organolanthanides to be synthesised and comprise three series: $Ln(C_5H_5)_3$, $Ln(C_5H_5)_2X$ and $Ln(C_5H_5)X_2$. They are prepared by the reaction of the lanthanide chloride with the stoichiometric amount of NaC_5H_5, followed, when appropriate, by metathesis:

$$LnCl_3 + nC_5H_5Na \xrightarrow{THF} LnCl_{3-n}(C_5H_5)_n + nNaCl$$

(i) Tricyclopentadienyls

These are volatile compounds, subliming *in vacuo* at 200–250°. Their solid-state structures display the effect of the lanthanide contraction (Figure 2.40). Thus in the lanthanum and praseodymium compounds, each metal is bound to one η^2 and three η^5-ligands; at the other extreme, in $Lu(C_5H_5)_3$, there are two η^5- and two η^1-rings. The erbium and thulium compounds contain isolated $(\eta^5\text{-}C_5H_5)_3Ln$ molecules.

$Ln(C_5H_5)_3$ are Lewis acids: with Lewis bases they form tetrahedral 1:1 adducts, $Ln(C_5H_5)_3.L$ (L = THF, NH_3, CH_3CN, $C_6H_{11}NC$, for example), and trigonal bipyramidal 2:1 adducts, $Ln(C_5H_5)_3(MeCN)_2$.

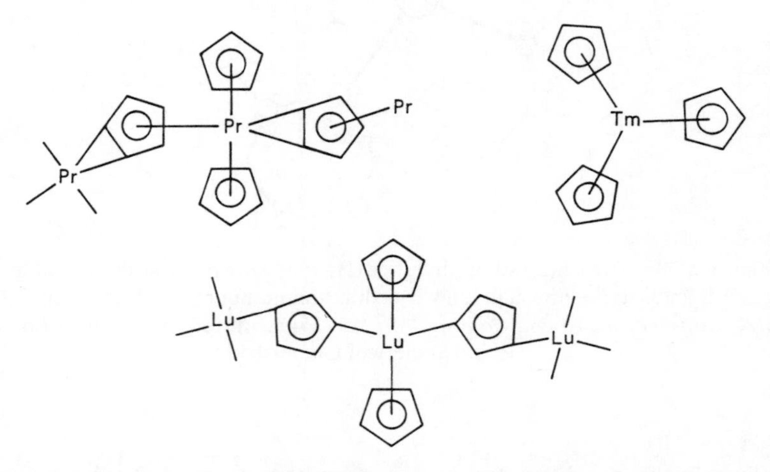

Figure 2.40 Structures of $(C_5H_5)_3Ln$ compounds showing the decrease in coordination number as the lanthanide becomes smaller

(ii) Dicyclopentadienyls

$Ln(C_5H_5)_2Cl$ are dimers (Figure 2.41) though the bridges are cleaved by donor solvents like THF. The chloride can be replaced by other ligands (such as H, CN, OMe, PBu^t_2, $OCOCH_3$, BH_4) as well as by alkyl groups:

$$Ln(C_5H_5)_2Cl \xrightarrow[\text{THF}]{\text{LiR}} Ln(C_5H_5)_2R$$

$$Ln(C_5H_5)_2Cl \xrightarrow{\text{LiAlMe}_4} (C_5H_5)_2Ln(\mu\text{-Me})_2AlMe_2 \xrightarrow{\text{Py}} (C_5H_5)_2Ln(\mu\text{-Me})_2Ln(C_5H_5)_2$$

These can be isolated as monomeric THF adducts $Ln(C_5H_5)_2R.THF$ or as dimeric $(C_5H_5)_2LnR_2Ln(C_5H_5)_2$. Some of these undergo interesting reac-

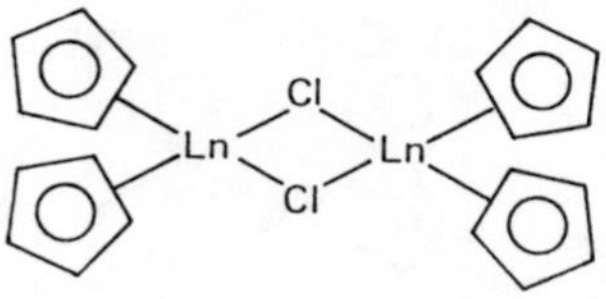

Figure 2.41

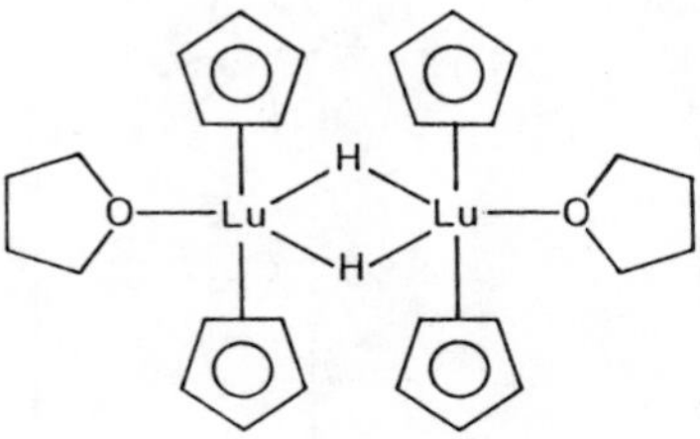

Figure 2.42

tions. Reduction of $Lu(C_5H_5)_2Cl$ with NaH in THF affords a dimeric hydride (Figure 2.42) which can also be made by reduction of some alkyls $Lu(C_5H_5)_2R$. The hydride adds on hex-1-ene and phenylacetylene to form hexyl and alkenyl derivatives.

(iii) Monocyclopentadienyls

These have been obtained for most lanthanides as tris(THF) solvates; the structure of the erbium compound is shown in Figure 2.43.

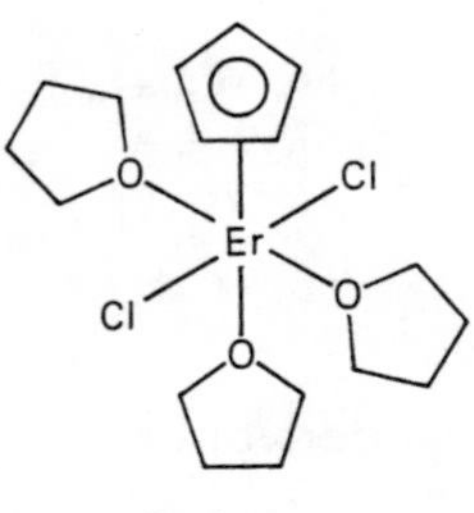

Figure 2.43

(iv) Substituted cyclopentadienyls

The most important are those of the pentamethylcyclopentadienyl ligand. Its increased bulk means that no more than two rings can coordinate to a lanthanide atom. Some of the fascinating chemistry observed with lutetium alkyls is shown in Figure 2.44.

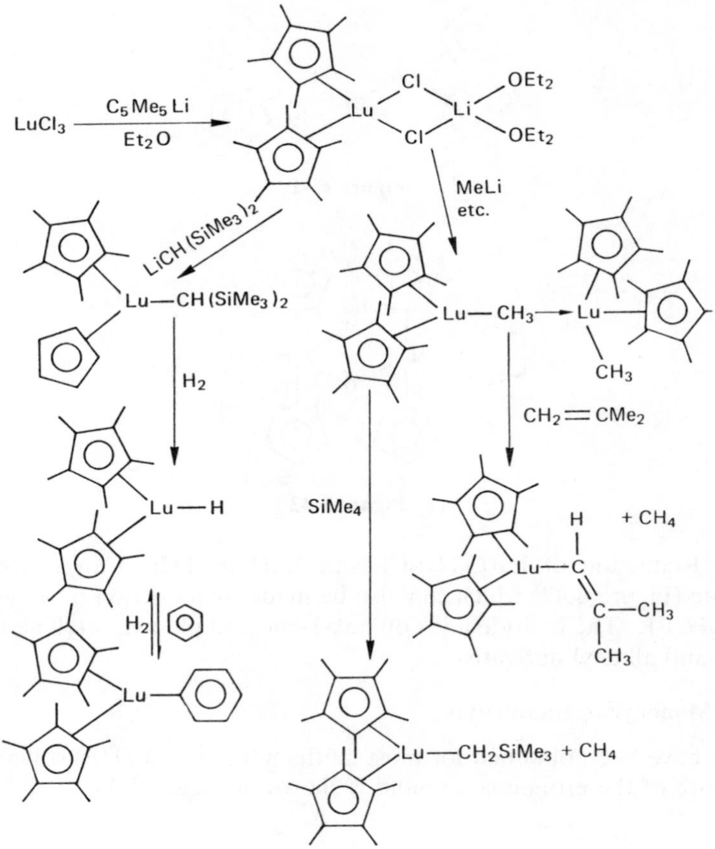

Figure 2.44

2.14.3 Cyclooctatetraenyl derivatives

Three types of compounds have been made: $KLn(C_8H_8)_2$, $Ln_2(C_8H_8)_3.(THF)_2$ and $[(C_8H_8)LnCl(THF)_2]_2$; the first two types contain the 'sandwich' ion $Ln(C_8H_8)_2^-$. Syntheses are:

$$LnCl_3 \xrightarrow[K_2C_8H_8]{1\ mol} [Ln(C_8H_8)Cl(THF)_2]_2 \xrightarrow{exc.K_2C_8H_8} K^+Ln(C_8H_8)_2^-$$

$$\text{Ln atoms} + C_8H_8 \xrightarrow[\text{2. THF}]{\text{1. Co-condense} \quad -196^\circ} [C_8H_8Ln(THF)_2]^+Ln(C_8H_8)_2^-$$

In the latter compounds, there is a metal–ring interaction that ensures coordinative saturation (Figure 2.45). The bonding in these compounds appears to be largely electrostatic, in contrast to uranocene; for example, they react immediately with UCl_4 to form $U(C_8H_8)_2$.

'Mixed' sandwich compounds $C_5H_5LnC_8H_8$ have also been made.

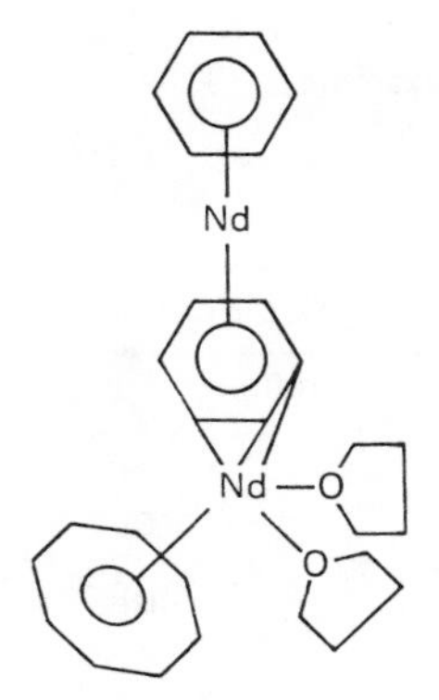

Figure 2.45

2.14.4 Arene complexes

Recent research has provided the first examples of such compounds, whose stability would not be expected on a purely ionic model (Figure 2.46; see page 82), through the use of good π-donors as ligands.

The 'sandwich' compounds of several lanthanides are stable at room temperature and have notable colours (for example, Lu–red-green; Y, Pr, Gd – purple).

2.14.5 Carbonyls

These are made by co-condensing lanthanide vapours with CO/Ar mixtures at 8–12K. Controlled heating affords the carbonyls $Ln(CO)_n$ (n = 1–6; Ln = Pr, Nd, Eu, Gd, Ho, Yb). They are only stable to around 40K but have been characterised by their IR spectra.

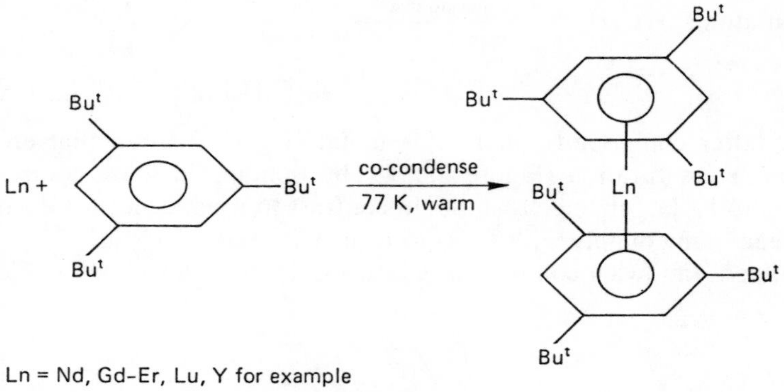

Ln = Nd, Gd–Er, Lu, Y for example

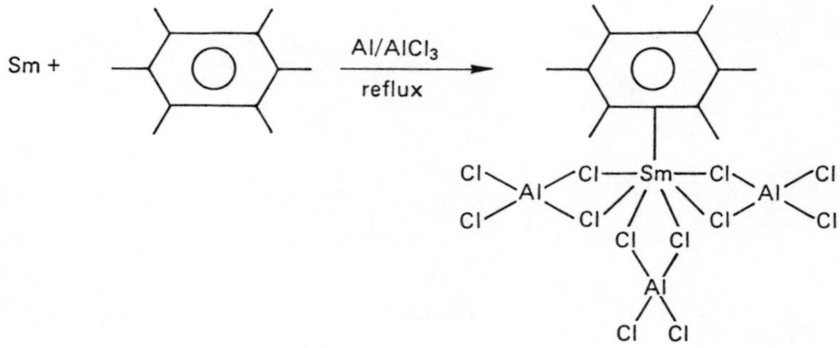

Figure 2.46

2.14.6 Compounds of lanthanides in the +4 oxidation state

A number of Ce(IV) compounds have been reported, but in view of the oxidising tendencies of Ce^{4+}, most are regarded as incompletely characterised.

$(C_8H_8)_2Ce$, isomorphous with $(C_8H_8)_2M$ (M = Th, U), is genuine, but others like $(C_5H_5)_4Ce$ and $(C_5H_5)_3CeX$ (X = alkyl, aryl, OR) are ill-defined, with the exception of the strongly oxidising $(C_5H_5)_3CeOCHMe_2$, volatile at 70° in a high vacuum.

2.14.7 Compounds of lanthanides in the +2 oxidation state

Simple cyclopentadienyls have been made for Eu, Yb and Sm, thus:

$$SmI_2 \xrightarrow[THF]{C_5H_5Na} (C_5H_5)_2Sm(THF)_2$$

$$(MeC_5H_4)_2Yb.CH_3 \xrightarrow[h\nu]{\text{toluene}} (MeC_5H_4)_2Yb$$

The most important compounds are those of the pentamethyl cyclopentadienyl ligand:

$$YbCl_2 \text{ or } EuCl_3 \xrightarrow[\text{THF}]{C_5Me_5Na} (C_5Me_5)_2M(THF)$$

$$\text{Sm atoms} \xrightarrow[\text{(ii) THF}]{\text{(i) } C_5Me_5H/\text{hexane } -110^\circ} (C_5Me_5)_2Sm(THF)_2$$

The adducts have 'bent' structures as have, surprisingly, the solvent-free $(C_5Me_5)_2M$ (M = Sm, Eu) (Figure 2.47), prepared by vacuum desolvation of the THF adducts. Remarkably, $(C_5Me_5)_2Sm$ reversibly complexes with nitrogen, forming a dimeric compound with 'side-on' bridging nitrogen (Figure 2.48).

2.14.8 Reactions of coordinated ligands in organometallics

In recent years it has become evident that organolanthanides can bring about unusual reactions. Some of these are shown in Figure 2.49, demonstrating a range of insertion reactions of carbon monoxide, including C–O bond breaking.

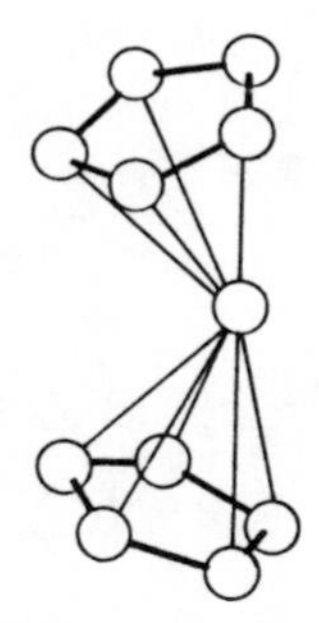

Figure 2.47 The 'bent sandwich' structure adopted by $(C_5Me_5)_2M$ (M = Eu, Sm)

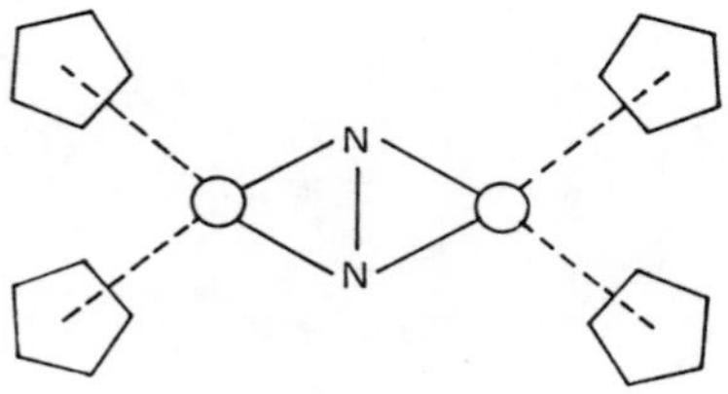

Figure 2.48

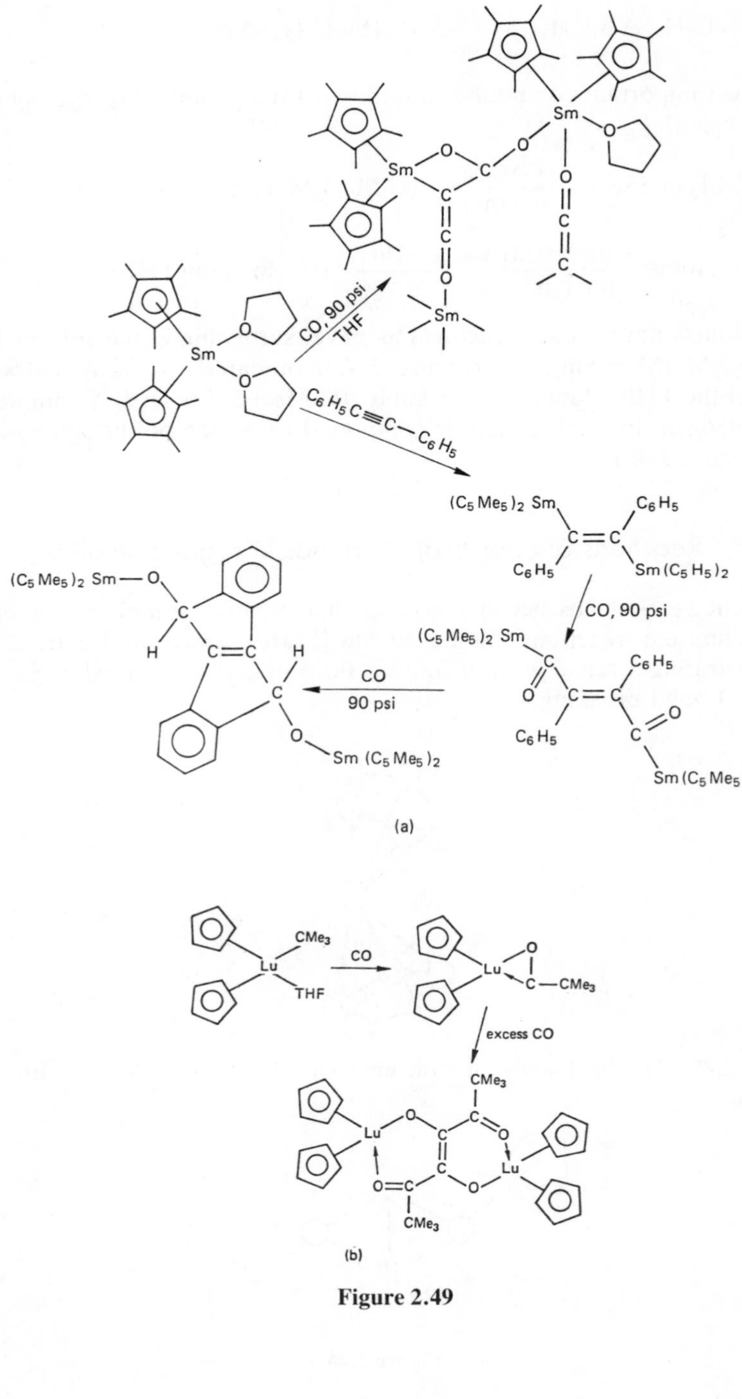

Sm
O
C
O
Sm
O
C
C
O
Sm
CO, 90 psi
THF
$C_6H_5C{\equiv}C{-}C_6H_5$
$(C_5Me_5)_2$ Sm
C_6H_5
C=C
C_6H_5
Sm $(C_5H_5)_2$
CO, 90 psi
$(C_5Me_5)_2$ Sm
C
O
C=C
C_6H_5
C_6H_5
C
O
Sm $(C_5Me_5)_2$
CO
90 psi
$(C_5Me_5)_2$ Sm—O
C
H
C=C
H
C
O
Sm $(C_5Me_5)_2$

(a)

CMe_3
Lu
THF
CO
Lu
O
C—CMe_3
excess CO
CMe_3
Lu
O
C
C
O
C
C
O
Lu
O=C
CMe_3

(b)

Figure 2.49

3
The actinides

3.1 Position in the Periodic Table

After the discovery of uranium – recognised in 1789 by Klaproth – and thorium (Berzelius, 1828), a considerable period elapsed before actinium (1899) and protactinium (1913) were recognised. For many years they were regarded as a new transition-metal series and placed in groups IIIA–VIA of the Periodic Table. Following the discovery of the transuranium elements, increased chemical and spectroscopic study, particularly of the later actinides with their increasing emphasis on the +3 state, revealed them to be a 'heavier version' of the lanthanide series, in which the $5f$ orbitals were being filled.

3.2 Synthesis of the actinides

Thorium and uranium are, of course, obtained from natural ores (section 3.5). While actinium and protactinium are present in ores, it is best to obtain actinium by neutron bombardment of lighter nuclei, and protactinium can also be obtained in this way:

$$^{226}\text{Ra} \xrightarrow{(n,\,\gamma)} {}^{227}\text{Ra} \longrightarrow {}^{227}\text{Ac} + \beta^-$$

$$^{230}\text{Th} \xrightarrow{(n,\,\gamma)} {}^{231}\text{Th} \longrightarrow {}^{231}\text{Pa} + \beta^-$$

(In fact, most of the world's supply of protactinium has been obtained by one large-scale work-up of uranium residues, section 3.5.)

The neutron-irradiation route is the general method used for transuranium elements, again by continuous irradiation in a nuclear reactor:

$$^{238}\text{U} \xrightarrow{(n,\,2n)} {}^{237}\text{U} \xrightarrow[7\ \text{d}]{\beta^-} {}^{237}\text{Np}$$

$$^{238}\text{U} \xrightarrow{(n,\,\gamma)} {}^{239}\text{U} \xrightarrow[23\ \text{min}]{\beta^-} {}^{239}\text{Np} \xrightarrow[2\ \text{d}]{\beta^-} {}^{239}\text{Pu}$$

The neutrons arise, of course, from fission of ^{235}U. ^{239}Pu in particular has been synthesised on the ton scale.

Synthesis of the elements beyond Pu requires successive neutron capture, so that yields are obviously less, even with a high-density flux of neutrons. Starting from 1 kg of ^{239}Pu using a neutron flux of 3×10^{14} neutrons cm^{-2} s^{-1} (a reasonably high value), around 1 mg of ^{252}Cf is obtained after 5–10 years. Some syntheses carried out by this route follow:

$$^{239}\text{Pu} \xrightarrow{n} {}^{240}\text{Pu} \xrightarrow{n} {}^{241}\text{Pu} \xrightarrow{n} {}^{242}\text{Pu} \xrightarrow{n} {}^{243}\text{Pu} \xrightarrow{n} {}^{243}\text{Am} \xrightarrow{n} {}^{244}\text{Am}$$

$$^{241}\text{Pu} \xrightarrow[14.4\ \text{y}]{\beta^-} {}^{241}\text{Am} \xrightarrow{n,\gamma} {}^{242}\text{Am} \xrightarrow[16\ \text{h}]{\beta^-} {}^{242}\text{Cm}$$

$$^{244}\text{Am} \xrightarrow[10\ \text{h}]{\beta^-} {}^{244}\text{Cm}$$

$$^{242}\text{Pu or } {}^{243}\text{Am or } {}^{244}\text{Cm} \xrightarrow{\text{multiple } n,\ \beta^-} {}^{249}\text{Bk},\ {}^{252}\text{Cf},\ {}^{253}\text{Es},\ {}^{257}\text{Fm}$$

This process reaches a limit at ^{257}Fm; the product of the next neutron capture undergoes spontaneous fission (^{258}Fm, $t_{1/2} = 0.38$ ms).

To reach elements beyond ^{257}Fm, two processes are theoretically feasible. One route uses an extremely intense neutron flux, such as encountered in a thermonuclear explosion, so that heavier nuclei are attained before the ^{258}Fm decays. Although this route has not been utilised to get beyond ^{257}Fm (and for socio-political reasons it is unlikely to happen in the future), it was used in the first synthesis of ^{255}Fm at Eniwetok in 1952:

$$^{238}\text{U} \xrightarrow{17n} {}^{255}\text{U} \xrightarrow{\beta^-} {}^{255}\text{Np} \xrightarrow{\beta^-} {}^{255}\text{Pu} \xrightarrow{\beta^-} {}^{255}\text{Am} \xrightarrow{\beta^-} {}^{255}\text{Cm}$$

$$^{255}\text{Cm} \xrightarrow{\beta^-} {}^{255}\text{Bk} \xrightarrow{\beta^-} {}^{255}\text{Cf} \xrightarrow{\beta^-} {}^{255}\text{Es} \xrightarrow{\beta^-} {}^{255}\text{Fm}$$

The extremely rapid multiple neutron capture during the thermonuclear explosion was followed by a series of rapid decays that afforded the product fermium.

The second route to heavy nuclei involves heavy-ion bombardment, accelerating suitable projectile nuclei (^{1}H, ^{11}B, ^{12}C, ^{16}O are common, but nuclei up to around ^{56}Fe are used) into actinide targets. Although reliable,

Table 3.1 The longest-lived actinide isotopes (* = natural)

Element	Mass of isotope	Half-life	Means of decay
Actinium	227*	21.77 y	β^-
Thorium	232*	1.41×10^{10} y	α
Protactinium	231*	3.28×10^4 y	α
Uranium	234*	2.45×10^5 y	α
	235*	7.037×10^8 y	α
	238*	4.47×10^9 y	α
Neptunium	236	1.55×10^5 y	β^-, EC
	237	2.14×10^6 y	α
Plutonium	239	2.41×10^4 y	α
	240	6.563×10^3 y	α
	242	3.76×10^5 y	α
	244	8.26×10^7 y	α
Americum	241	432.7 y	α
	243	7.38×10^3 y	α
Curium	244	18.11 y	α
	245	8.5×10^3 y	α
	246	4.73×10^3 y	α
	247	1.56×10^7 y	α
	248	3.4×10^5 y	α
	250	$\sim 1.13 \times 10^4$ y	SF
Berkelium	247	1.38×10^3 y	α
	249	320 d	β^-
Californium	249	351 y	α
	250	13.1 y	α
	251	898 y	α
	252	2.64 y	α
Einsteinium	252	472 d	α
	253	20.47 d	α
	254	276 d	α
	255	39.8 d	β^-
Fermium	257	100.4 d	α
Mendelevium	258	55 d	α
Nobelium	259	1 h	α, EC
Laurencium	260	3.0 min.	α

this route has the drawbacks of low yields (the products are produced an atom at a time) and the availability of suitable actinide targets. Some syntheses follow:

$$^{244}\mathrm{Cm} \xrightarrow{\alpha,\ ^1_1\mathrm{H}} {}^{247}\mathrm{Bk}$$

$$^{253}\mathrm{Es} \xrightarrow{\alpha,\ ^1_0n} {}^{256}\mathrm{Md}$$

$$^{248}\mathrm{Cm} + {}^{18}\mathrm{O} \longrightarrow {}^{259}\mathrm{No} + {}^{4}\mathrm{He} + 3^1_0n$$

$$^{249}\mathrm{Cf} + {}^{12}\mathrm{C} \longrightarrow {}^{255}\mathrm{No} + {}^{4}\mathrm{He} + 2^1_0n$$

$$^{248}\mathrm{Cm} + {}^{15}\mathrm{N} \longrightarrow {}^{260}\mathrm{Lr} + 3^1_0n$$

$$^{249}\mathrm{Cf} + {}^{11}\mathrm{B} \longrightarrow {}^{256}\mathrm{Lr} + 4^1_0n$$

3.3 Oxidation states and electron configurations

The trend in oxidation states across the actinide series is represented in Figure 3.1; it shows the oxidation states that have been characterised as reasonably stable in aqueous solution or in solid compounds.

Except for thorium and protactinium all the actinides form An^{3+} ions in solution (and these two 'exceptions' form solid compounds in the +3 state). To this extent only does the actinide series resemble the lanthanides, for not only is there a stable +4 state for over half the 5*f* metals but several elements exhibit a stable +6 or +7 state. The ability of the early actinides to form compounds in such high oxidation states resembles the *d*

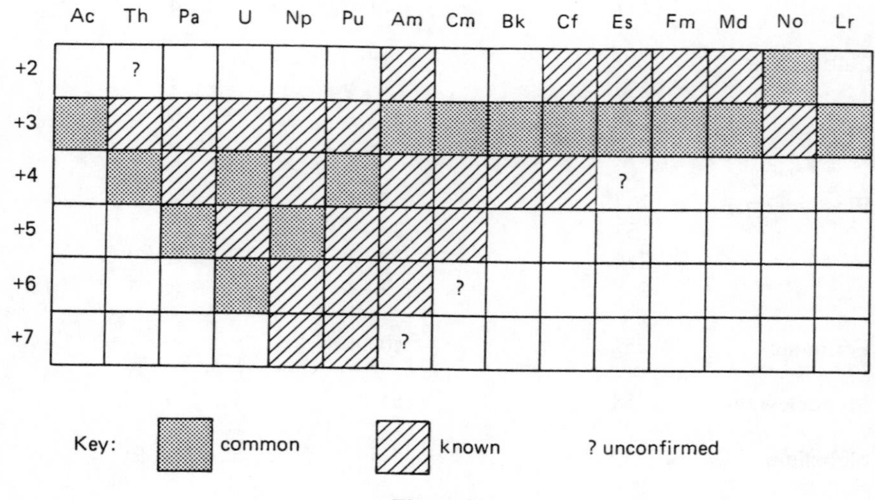

Figure 3.1

block metals, and must reflect corresponding low ionisation energies (the first four ionisation energies of Th are 587, 1110, 1978 and 2780 kJ mol^{-1}; compare the corresponding values for Zr: 660, 1267, 2218, 3313 kJ mol^{-1}) and is a consequence of the near-degeneracy of the $7s$, $6d$ and $5f$ electrons. (In the lanthanide series, the $4f$–$5d$ gap is much bigger.) This is reflected in the electron configurations of the actinides, shown in Figure 3.2.

It will be noted that, at thorium, the $6d$ electrons are lower in energy than those in the $5f$ orbitals, but that removal of the $7s$ electrons in ion formation stabilises the $5f$ energy levels. Furthermore, on crossing the actinide series the $5f$ shell becomes stabler as electrons are added to it, presumably for similar reason to those applying to the d-block transition metals.

Examination of the electrode potentials (reduction potentials) (Table 3.2) enables much of the chemistry of these elements to be rationalised.

Study of the $M^{3+}-M^{2+}$ potentials reveals the increasing stability of the +2 state in the second half of the series, an effect much more noticeable than in the lanthanide series. Dihalides MX_2 (X = Cl, Br, I) have actually been isolated for M = Am, Cf, Es, but not for the intervening Cm and Bk (which are obviously less stable). It would be predicted that such dihalides would be isolable for the actinides after Es, with the possible exception of Lr, but the short half-lives and experimental difficulties in studying the chemistry of the later actinides make confirmation difficult. There is, however, some evidence for $FmCl_2$.

	Atom	M^{3+}	M^{4+}
Ac	$6d^1\ 7s^2$	—	—
Th	$6d^2\ 7s^2$	$5f^1$	—
Pa	$5f^2\ 6d^1\ 7s^2$	$5f^2$	$5f^1$
U	$5f^3\ 6d^1\ 7s^2$	$5f^3$	$5f^2$
Np	$5f^4\ 6d^1\ 7s^2$	$5f^4$	$5f^3$
Pu	$5f^6\ 7s^2$	$5f^5$	$5f^4$
Am	$5f^7\ 7s^2$	$5f^6$	$5f^5$
Cm	$5f^7\ 6d^1\ 7s^2$	$5f^7$	$5f^6$
Bk	$5f^9\ 7s^2$	$5f^8$	$5f^7$
Cf	$5f^{10}\ 7s^2$	$5f^9$	$5f^8$
Es	$5f^{11}\ 7s^2$	$5f^{10}$	$5f^9$
Fm	$5f^{12}\ 7s^2$	$5f^{11}$	$5f^{10}$
Md*	$5f^{13}\ 7s^2$	$5f^{12}$	$5f^{11}$
No*	$5f^{14}\ 7s^2$	$5f^{13}$	$5f^{12}$
Lr*	$5f^{14}\ 6d\ 7s^2$	$5f^{14}$	$5f^{13}$

Configurations given are those outside a radon 'core':
$1s^2\ 2s^2\ 2p^6\ 3s^2\ 3p^6\ 3d^{10}\ 4s^2\ 4p^6\ 4d^{10}\ 4f^{14}\ 5s^2\ 5f^6\ 5d^{10}\ 6s^2\ 6p^6$

*Predicted.

Figure 3.2 Electron configurations of the actinides

Table 3.2 Reduction potentials for the actinide (4+) and (3+) ions

	E^0(V) for														
	Ac	Th	Pa	U	Np	Pu	Am	Cm	Bk	Cf	Es	Fm	Md	No	Lr
$M^{3+} + e^- \rightarrow M^{2+}$	−4.9	−4.9	−4.7	−4.7	−4.7	−3.5	−2.3	−3.7	−2.8	−1.6	−1.6	−1.1	−0.15	+1.45	
$M^{4+} + e^- \rightarrow M^{3+}$		−3.7	−2.0	−0.63	0.15	0.98	2.3	3.1	1.64	3.2	4.5	4.9	5.4	6.5	7.9

Data from L. J. Nugent in *MTP International Review of Science and Technology*, *Series 2*, Butterworths, 1975, p. 195.

Turning to the $M^{4+}-M^{3+}$ potentials, again we observe correlation with chemical evidence, using these potentials as a rough indication of the stability of the +3 and +4 ions. The potentials suggest that there is a trend to decreasing stability of the +4 state (notice the discontinuity for Bk^{4+} with its half-filled shell). Thus ThI_3 and U^{3+}(aq) (and other U(III) compounds) reduce water, the former rapidly, but the following elements all have +3 ions that are stable in water, Np^{3+} in the absence of air, the succeeding actinides even in the presence of oxygen.

Alternatively, considering the +4 state, it is effectively the only one for thorium, and quite stable for Pa^{4+}, U^{4+} and Np^{4+}. Pu^{4+} tends to disproportionate, Am^{4+} and Cm^{4+} only exist in solution as fluoride complexes (in the latter case, unstable ones), but there is an increase in stability for Bk^{4+} owing to the change in potential consequent on the low I_4 on forming the f^7 Bk^{4+} ion. Tetrafluorides have been made for all the metals from Th to Cf; in view of the standard potential for fluorine ($F_2 + 2e^- \rightarrow 2F^-$; $E^0 = 2.85$ V), this may represent the feasible limit (there have been claims for EsF_4). The tetrachlorides are more limited (the standard potential for chlorine is 1.36 V), being found for Th–Np, though $PuCl_4$ exists in the gas phase at high temperatures.

The lack of compounds in high oxidation states late in the series thus reflects the reduction potentials, in turn a consequence of the fact that the 4th ionisation energy for the later actinides is too large to be balanced by any gain in bonding energies.

3.4 Characteristics of the actinides

Whereas the lanthanides form a series of metals with very similar properties, largely displaying a smooth trend from one end of the series to the other, the actinides are more complex.

Essentially, the early actinides most resemble the early *d*-block metals with a tendency to variability in oxidation state and the ability to form stable complexes with ligands like chloride, sulphate, carbonate and acetate. There is strong evidence for covalent contribution to the bonding in ions like MO_2^{2+} and in compounds with π-bonding ligands, such as $U(C_8H_8)_2$.

In mid-series around Bk, the trend towards lanthanide-like behaviour becomes marked, with a preference to the 3+ oxidation state. This may be associated with the steady rise in ionisation potentials across the series (page 170).

3.5 Extraction of the metals

3.5.1 The extraction of thorium

The principal thorium ore is monazite sand, important producing areas being India (Travancore), Brazil, Ceylon, the USSR and the USA (Florida, Idaho and the Carolinas). Monazite is a rare earth orthophosphate ($Ce/Nd/LaPO_4$) in which a proportion of the lanthanides are replaced by thorium (up to 20 per cent of lanthanide content). Of course, the ore is also worked to extract lanthanides.

Thorium can be extracted by one of two routes, one involving digestion with acid (H_2SO_4), the other alkali digestion (NaOH). In the acid process, digestion for 3–4 hours at around 230° will digest some 99 per cent of the ore. Ammonia solution is then added to raise the pH to 1, at which point thorium phosphate precipitates but most of the lanthanides remain in solution. Treatment of the thorium phosphate with alkali turns it into thorium hydroxide (phosphate and any sulphate passing into solution), which is redissolved in nitric acid. The resulting thorium nitrate can be purified by extraction into kerosene using tributylphosphate, separating it from any remaining lanthanides. (For the alkali route, see section 2.2.)

3.5.2 The extraction of protactinium

Protactinium has generally been obtained as a by-product from uranium extraction. Most of the world's Pa reserves were obtained some 30 years ago in the UK from about 60 tons of pitchblende residues, the so-called 'ethereal sludge', which contained about 12 t of uranium, the protactinium being precipitated in the ether extraction plant on account of the low acidity of the nitric acid solution (see section 3.14).

The residues were first leached with 4M HNO_3(0.1M in HF) to remove U and Pa, the uranium being separated by extraction with tributylphosphate. Most of the protactinium was precipitated by the addition of $AlCl_3$; NaOH digestion of the precipitate removed silicon, aluminium and phosphate. The resulting solid was dissolved in 8M HCl (containing 0.1M HF to ensure Pa complexation again) and purified by repeated extraction with disobutylketone and stripping with the HCl/HF solution. Eventually a solution containing 126.75 g Pa was obtained.

3.5.3 Uranium ores and extraction of uranium

Uranium is more widespread in the earth's crust (2 ppm) than a number of 'familiar' elements, such as silver, but unfortunately more widely dispersed. Some 200 ores contain uranium, but only a small number of these are commercially important. The most abundant are uraninite and pitchblende, which are oxide ores with variable concentration in the region UO_2–$UO_{2.67}$. Others include:

autunite $Ca(UO_2)_2(PO_4)_2.nH_2O$ $(n = 8–12)$

torbernite $Cu(UO_2)_2(VO_4)_2.nH_2O$ $(n = 8–12)$

carnotite $K_2(UO_2)_2(VO_4)_2.3H_2O$

and uranophane $Ca(UO_2)_2Si_2O_7.6H_2O$

It also occurs in a number of rare earth ores (euxenite, fergusonite, sarmarskite, among others).

Most of the readily accessible ores are found in Canada, the USA (Colorado and Wyoming), South Africa (Rand goldfields), Australia and France, though rich veins of ore have been found in low total abundance elsewhere. Uranium is found in many parts of the world in lower-abundance ores which are not at present economically mineable; there is calculated to be some 4.5×10^9 t of uranium in the world's oceans (at an average concentration of 3.3 mg m^{-3}) and interest has been expressed in its recovery.

The uranium ores are pre-concentrated (if possible), one method being the somewhat improbable mechanical sorting on the basis of a radiation reading for each lump of ore. This is followed by extraction of the uranium, either by ion-exchange or solvent extraction.

Acid leaching uses sulphuric acid and an oxidising agent such as MnO_2 or $NaClO_3$ (to oxidise the uranium present to the extractable U(VI) so that the uranium is extracted as a sulphate complex, such as $UO_2(SO_4)_3^{4-}$). Alkali leaching utilises sodium carbonate together with oxygen as the oxidant to extract the uranium first by absorption of the complexes on anion-exchange resins and then elution with, for example, 1M NaCl; solvent extraction uses extractants such as tertiary amines in kerosene solution.

The product of the foregoing process is an acidic solution of a uranyl complex of ligands such as chloride and sulphate; on neutralisation, usually with ammonia, a precipitate of 'yellow cake' (analysing as $(NH_4)_2U_2O_7$) is obtained. Alternatively, uranium is precipitated as the peroxide $UO_4.xH_2O$. On ignition, 'yellow cake' affords an oxide of uranium. These products are generally purified further, usually via solvent extraction, to afford nuclear-grade material.

Pure uranium can be obtained by the following process:

1 Uranium ore concentrates are purified after digestion by nitric acid and finally with solvent extraction using tributylphosphate in kerosene to afford uranyl nitrate:

$$U_3O_8 + 8HNO_3 \longrightarrow 3UO_2(NO_3)_2 + 2NO_2 + 4H_2O$$

2 The nitrate undergoes thermal decomposition:

$$UO_2(NO_3)_2.6H_2O \longrightarrow UO_3 + 2NO_2 + \tfrac{1}{2}O_2 + 6H_2O$$

3 Uranium trioxide is reduced to the dioxide:

$$UO_3 + H_2 \longrightarrow UO_2 + H_2O$$

4 Uranium dioxide is converted to UF_4, 'Green salt':

$$UO_2 + 4HF \longrightarrow UF_4 + 2H_2O$$

5 The UF_4 undergoes metallothermic reduction with magnesium:

$$UF_4 + 2Mg \longrightarrow U + 2MgF_2$$

3.6 Isotope separation methods

Separating the fissionable ^{235}U needed for nuclear reactors from the bulk ^{238}U isotope has been accomplished in a number of ways, particularly gaseous diffusion, electromagnetic methods, gas centrifugation and laser separation. A brief description of each follows.

(a) *Gaseous diffusion*. The uranium is converted to UF_6 and allowed to diffuse at 70–80° through barriers of metals such as nickel or aluminium with a pore size around 10–25 nm. Following Graham's Law, the separation factor, α^*, is defined:

$$\alpha^* = \sqrt{\left(\frac{M_r(^{238}UF_6)}{M_r(^{235}UF_6)}\right)} = \sqrt{\left(\frac{352}{349}\right)} = 1.00429$$

By using a 'cascade' process with up to 3000 stages, an enrichment to 90 percent ^{235}U is obtained.

The process requires the use of fluorinated materials for valves, lubricants and fluorine-resistant metals, as well as scrupulous prevention of leaks and ingress of water. The barrier materials have to be manufactured to high tolerances to ensure consistent pore size; large-scale separation requires a large-scale (acres) total barrier size. The energy requirement to 'pump' the UF_6 is considerable.

(b) *Electromagnetic separation*. This is historically important, as it was the means by which isotopes were separated for the Manhattan project. The separation was achieved using cyclotron-like systems, with an input of UCl_4.

(c) *Laser separation*. This has been achieved in two ways.

One route uses uranium vapour and selective photoionisation: the vapour is excited simultaneously by UV light and a beam of light from a tuneable laser. Selective ionisation turns the ^{235}U atoms into $^{235}U^+$ ions which can be collected on a negative electrode.

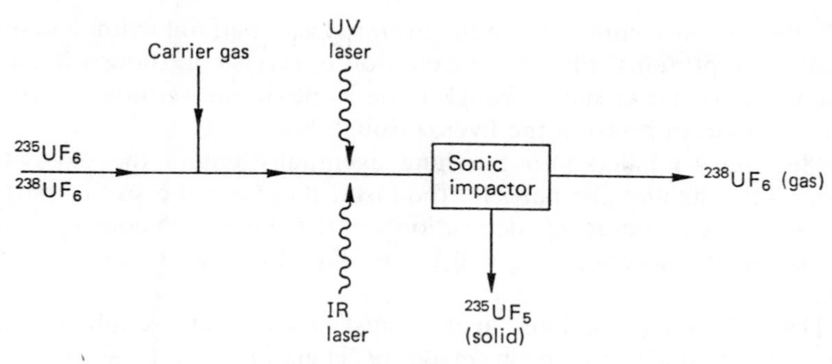

Figure 3.3

The second route uses the vapour of a volatile compound such as UF_6. Thus the U–F stretching vibration in $^{235}UF_6$ is selectively excited with a pulse of energy from a tuneable IR laser (supercooled vapour is used in order to obtain sharp bands from ground-state UF_6 molecules as the isotope shift is of the order of 0.5 cm^{-1}). On irradiation with a high-intensity UV laser, the $^{235}UF_6$ decomposes into $^{235}UF_5$, while the $^{238}UF_6$ is unaffected; the resulting solid $^{235}UF_5$ is separated from the gaseous $^{238}UF_6$ in a sonic impactor (Figure 3.3).

The volatile alkoxide $U(OMe)_6$ has recently attracted attention because of its potential in laser isotope separation.

(d) *Gas centrifuge*. The principle of this method relies on the fact that if a volatile U compound such as UF_6 is centrifuged in a gas centrifuge, $^{238}UF_6$ tends to predominate in the periphery with $^{235}UF_6$ predominating in the axial position. The separation improves at higher speeds and lower temperatures. Various centrifuge types have been developed and used in the USA, the UK, and certain European countries.

3.7 Toxicity of the actinides

Elements whose lethal doses are considered to be at the microgram level have to be treated with respect. Of the actinides, plutonium has been the subject of most study; like other long-lived α-emitters, whose radiation can be shielded by a thin sheet of paper, its greatest danger lies in ingestion and inhalation. Ingested plutonium first passes into body fluids like blood but tends to concentrate in the liver and bone, leading to neoplasm (1 mg of ^{239}Pu emits over a million α-particles a second). Potential treatments for plutonium poisoning should therefore try to remove plutonium from the system quickly while it is still in a soluble form, and also be capable of removing the remaining immobilised plutonium.

Pu^{4+} in the blood is mainly bound to transferrin (the iron transport agent). Possibly because of its similar charge/radius ratio, it tends to mimic

Fe^{3+}. Plutonium accumulates in the liver and is bound to ferritin and other iron-storage proteins. Plutonium excretion is very slow, though it can be passed out via the kidneys through urine as plutonium–citrate complexes or, less importantly, from the liver, through bile.

Plutonium inhaled into the lung eventually enters the circulatory system – the smaller the particle, the faster its entry; the size of particle also affects the point of deposition in the lung. Obviously, before dissolution, parts of the lung will be irradiated by the α-emitting plutonium.

The plutonium-binding agent must form water-soluble complexes – much study has been made of chelating agents such as the hexadentate EDTA, commonly used to treat heavy-metal poisoning. The most effective has been diethylene-triaminepentaacetic acid (DTPA), which is potentially octadentate thus capable of satisfying the 'normal'

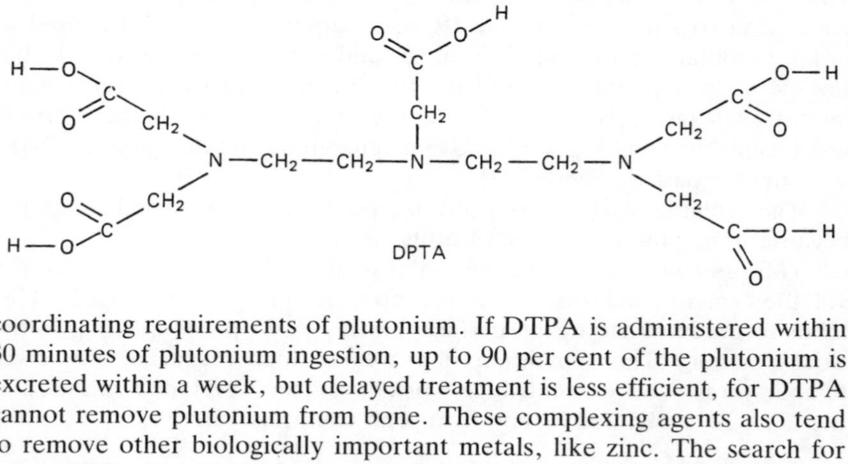

DPTA

coordinating requirements of plutonium. If DTPA is administered within 30 minutes of plutonium ingestion, up to 90 per cent of the plutonium is excreted within a week, but delayed treatment is less efficient, for DTPA cannot remove plutonium from bone. These complexing agents also tend to remove other biologically important metals, like zinc. The search for more efficient actinide binding agents has led to the study of catecholate complexes, based on the observed ability of the siderochromes, natural iron-binding catecholates. A molecule with four catecholate groups will again 'occupy' eight plutonium coordination sites. The rigid molecule (Figure 3.4) has a stability constant in excess of 10^{52} for its Pu^{4+} complex.

Flexible coordinating agents are even more efficient, particularly with anionic groups to improve water-stability; *in vivo* study in dogs and mice of the plutonium(IV) complex illustrated in Figure 3.5 shows it to be the most effective non-toxic Pu-chelating system yet studied.

3.8 Nuclear waste processing and disposal

The process of nuclear fission produces large amounts of energy, and also plutonium and transplutonium elements (from neutron-irradiation of ^{238}U)

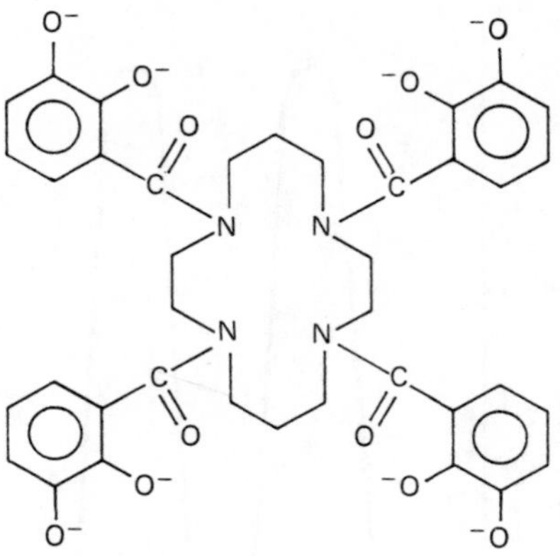

Figure 3.4

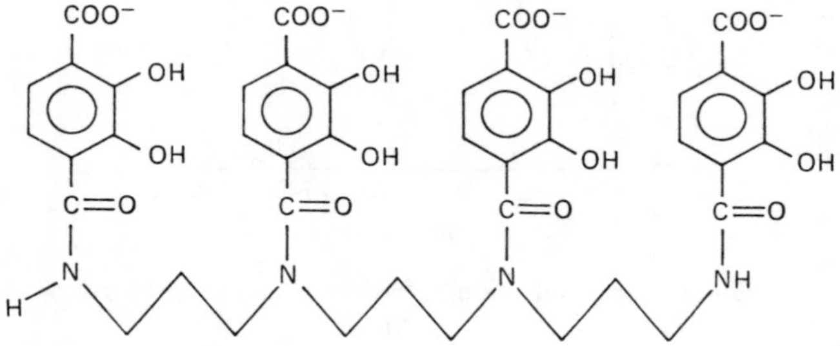

Figure 3.5

and fission products. Many of these fission products are efficient neutron-absorbers and thus prevent the fission reaction in the fuel elements running its full course. A nucleus of mass 235 or 239 on fission tends to produce two large but unequal nuclei, in the approximate mass ranges 90–100 and 130–145 (Figure 3.6).

The fission products thus consist principally of certain 4*d* transition metals and *s*-block metals as well as the lighter lanthanides; of these, the most important are probably ^{140}La, ^{141}Ce, ^{144}Pr, ^{95}Zr, ^{103}Ru and ^{95}Nb, as well as the long-lived ^{137}Cs, ^{90}Sr and ^{91}Y. Reprocessing seeks to separate the fission products and transplutonium elements, which present a toxic hazard but have no real use, from plutonium and uranium, which can be recycled as fuels.

There are essentially four steps in the separation process:

1 A 'cooling-off' period (up to 3 months) by immersion of the used fuel rods in ponds of water. This allows the more intensely radioactive but short-lived fission products (such as ^{131}I) to decay.

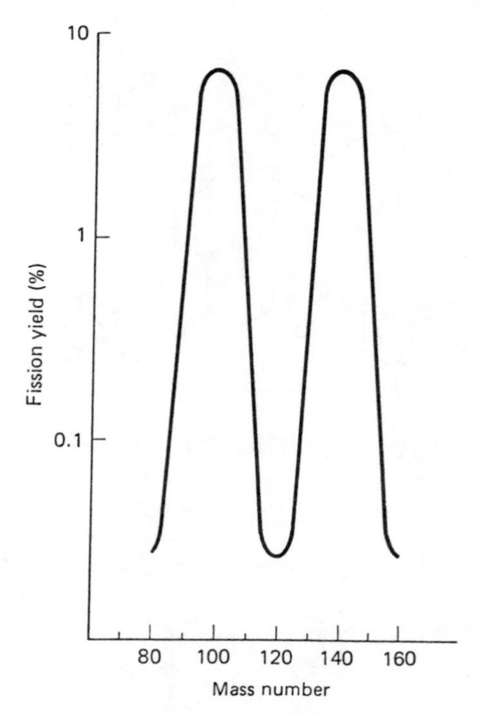

Figure 3.6 Schematic representation of fission product yield for fission of ^{235}U or ^{239}Pu

2 The fuel rods are dissolved in fairly concentrated (7M) HNO_3, affording a solution containing a mixture of principally $UO_2(NO_3)_2$, $Pu(NO_3)_4$ and other metal nitrates.

3 The solution is extracted with a solution of tributylphosphate (TBP) in kerosene. (Neat TBP is not used as it is too viscous and also has a very similar density to water.) This removes the uranium and plutonium by forming the neutral, kerosene-soluble, complexes $UO_2(NO_3)_2(TBP)_2$ and $Pu(NO_3)_4(TBP)_2$; the transplutonium metals and metals among the fission products form much weaker TBP complexes, which do not extract, but remain in the aqueous layer.

4 On reduction of the plutonium with a suitable reducing agent such as iron(II) sulphamate, hydrazine or hydroxylamine nitrates, the plutonium(III) passes into an aqueous phase as it no longer forms a strong TBP complex. Under these conditions, uranium(VI) is not reduced, and so remains in the kerosene layer.

$$UO_2^{2+} + 4H^+ + 2e^- \longrightarrow U^{4+} + 2H_2O \qquad E = +0.27\ \text{V}$$

$$Pu^{4+} + e^- \longrightarrow Pu^{3+} \qquad E = +1.00\ \text{V}$$

$$Fe^{3+} + e^- \longrightarrow Fe^{2+} \qquad E = +0.77\ \text{V}$$

After this, the uranium and plutonium are separately converted to their dioxides for re-use.

The problem of the fission products remains; they have to be stored in such a way as to prevent them escaping into the environment. At present, the 'preferred' process is immobilisation in a suitable geological repository. Prior to this, the evaporated waste solutions are heated to form a mixture of oxides, then mixed with chemicals such as silica and borax. A borosilicate glass is obtained on heating this mixture; this may be further encapsulated in an inert material (such as lead) before burial in a stable geological zone – rock salt and clays being favoured at the moment.

3.9 Some problems encountered in studying the actinides – and some solutions

These elements have presented a number of difficulties for chemists, who thus have needed to develop some special techniques.

The large quantities handled in the nuclear industry require both shielding and remote handling (not to mention care in ensuring that quantities below the critical mass are involved). The toxicity of the actinides is the subject of section 3.7. Even the small quantities involved in normal laboratory work are usually handled in glove boxes slightly below atmospheric pressure to reduce the likelihood of escape, as even the α-emitters are a real hazard if ingested.

Besides the 'direct' hazard to humans of the radiation, it can cause radiation damage to solutions, generating radicals (H · and OH ·) and H_2O_2, so that changes in oxidation state are likely; thus, high oxidation states are often reduced. Furthermore, heating problems are often encountered (^{242}Cm gives out some 122 W g^{-1}). Such problems are mitigated by the change to a longer-lived isotope (when available in sufficient quantity) – like ^{244}Cm ($t_{1/2}$ 17.6 y) instead of ^{242}Cm ($t_{1/2}$ 163 d).

The very small amounts available of the transamericium elements have meant handling microgram-quantities (or less), as follows. After ion-exchange purification (though cumulative contamination by daughter products can present problems over the length of an experiment), it is 'loaded' on to a single ion-exchange resin 'bead' (or small cube of charcoal). This is turned into the metal oxide at around 1000°, the oxide being confined to a quartz capillary attached to a high-vacuum system. All reactions are carried out directly in the capillary, which can be sealed for

X-ray studies; some case studies are described in section 3.19. Such ultramicro chemical studies also involve the use of microscopes to assist the manipulation of small quantities – both of solids and solutions – invisible to the naked eye. Although not applicable to the study of pure materials, or in determining spectroscopic properties, tracer studies involving a small quantity of an isotope with a large excess of a suitable non-radioactive carrier are frequently of value; by studying a precipitation reaction, for example, and measuring the radioactivity of the precipitate, it is possible to assess the solubility of the appropriate compound of the radioactive element.

Crystallography has its own difficulties. Quite apart from fogging of X-ray film, radioactive decay causes defects and dislocations in crystals; these are sometimes minimised by working at a temperature just below the melting point of the crystal, so that the defects are removed by annealing. In compounds containing heavy atoms, like uranium, the scattering due to the heavy atom tends to dominate that of light atoms (such as hydrogen) so that, ideally, complementary neutron-diffraction studies should be carried out to locate the light atoms.

Study of the elements with atomic numbers in excess of 100 has been complicated by the fact that they are produced one atom at a time. The combination of ultrasmall quantity and short half-life has meant the development of apparatus like that shown in Figure 3.7 for the study of element 104.

A brief description follows: 'new' atoms produced by the nuclear bombardment are ejected from the target, and carried by the aerosol of NaCl in helium to be deposited on the bottom surface of a polypropylene 'rabbit', which is pneumatically transported to a turntable (4). Here the sample can either be α-counted on a second turntable (4a) or dissolved (at 5a), chromatographed (6), evaporated (7–8) and α-counted (9–10). The whole system is controlled by a computer linked with an array of microswitches, conductivity and photoelectric sensors, so that a whole experiment can be completed within 3 minutes.

3.10 Systematics and trends in the actinide halides

Although the halides are generally treated separately, under the element in question, in this section they are used as a vehicle to systematise aspects of actinide chemistry.

Figure 3.8 lists the known actinide halides. Early in the series, the behaviour parallels that observed with the *d*-block metals, with the higher oxidation state equal to the number of 'outer-shell' electrons. This pattern ceases after uranium (there is no NpF_7 nor PuF_8); particularly after americium, the resemblance to the lanthanides is pronounced.

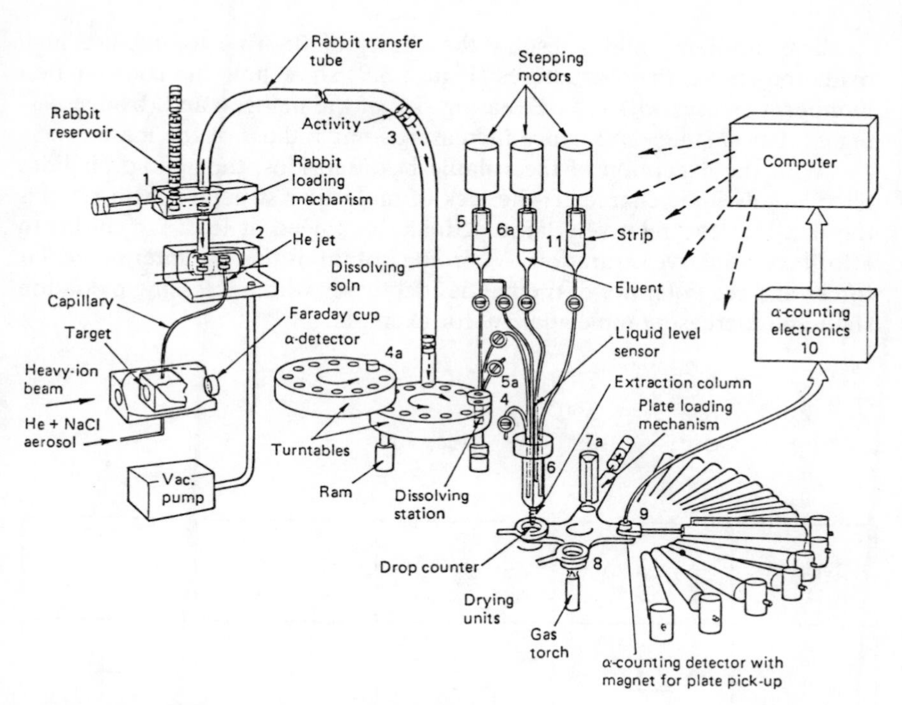

Figure 3.7 Computer-controlled apparatus used in the study of the chloride of element 104 [from E. K. Hulet *et al.*, *J. Inorg. Nucl. Chem.*, **42**, (1980) 79; reprinted with permission of Pergamon Press PLC © 1980]

	Oxidation state				
	+2	+3	+4	+5	+6
Ac		FClBrI			
Th	I	I	FClBrI		
Pa		I	FClBrI	FClBrI	
U		FClBrI	FClBrI	FClBr	FCl
Np		FClBrI	FClBrI	F	F
Pu		FClBrI	F		F
Am	ClBrI	FClBrI	F		F?
Cm		FClBrI	F		F?
Bk		FClBrI	F		
Cf	ClBrI	FClBrI	F		
Es	ClBrI	FClBrI	F?		
Fm					
Md					
No					
Lr					

Figure 3.8 The actinide halides

The uranium halides display the ability of fluorine to stabilise high oxidation states; their structures (Figure 3.9) show how the coordination number increases with (a) decreasing size of the halogen and (b) decreasing oxidation state (and hence increasing ionic radius) of the metal.

With the exception of the volatile hexafluorides, the actinide halides have largely ionic character; the lack of molecular structures also reflects the fact that the relatively large actinide ions need at least six anions to afford coordinative saturation. A corollary of this is that the intermolecular forces and the volatility of the halides decreases with decreasing oxidation state and increasing molecularity, for example:

	UCl_3	UCl_4	UCl_5	UCl_6
mp	835°	590°	—	177.5° (dec.)
bp	dec.	790°	dec. 300°	—

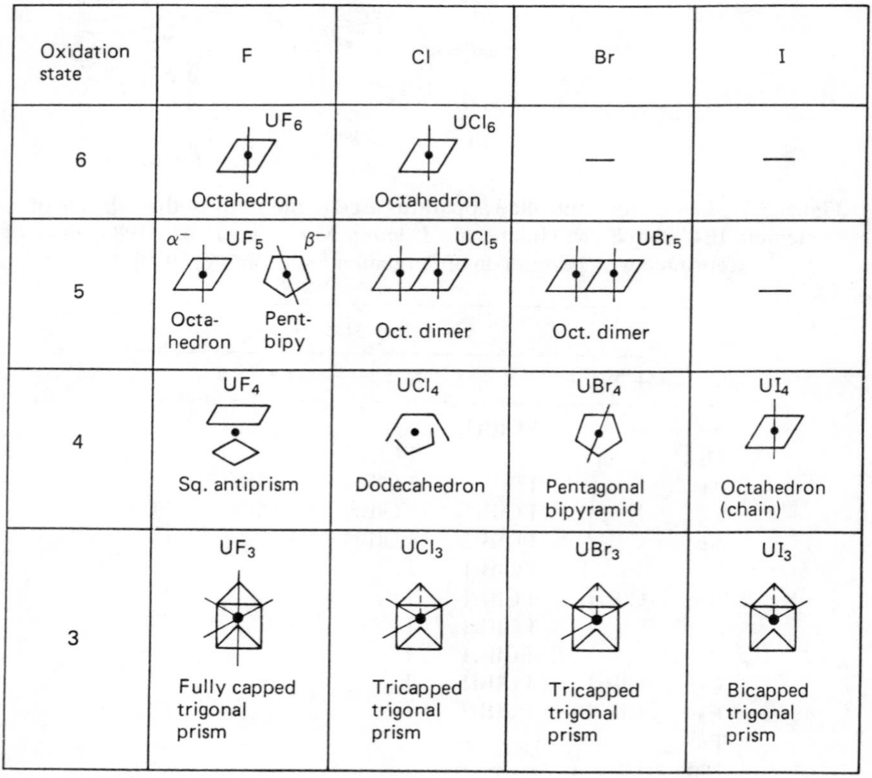

Oxidation state	F	Cl	Br	I
6	UF_6 Octahedron	UCl_6 Octahedron	—	—
5	α^- UF_5 β^- Octa-hedron Pent-bipy	UCl_5 Oct. dimer	UBr_5 Oct. dimer	—
4	UF_4 Sq. antiprism	UCl_4 Dodecahedron	UBr_4 Pentagonal bipyramid	UI_4 Octahedron (chain)
3	UF_3 Fully capped trigonal prism	UCl_3 Tricapped trigonal prism	UBr_3 Tricapped trigonal prism	UI_3 Bicapped trigonal prism

Figure 3.9 Coordination polyhedra in the uranium halides [adapted from J. C. Taylor, *Coord. Chem. Revs*, **20** (1976) 203; reprinted with permission of Elsevier Science publishers]

Synthetic routes

There are many preparative paths to the halides; some of these are discussed in the section on the trihalides. In general, direct reaction with the halogen affords a high available oxidation state, whereas reaction with hydrogen halides often leads to lower oxidation states, as do thermal decomposition routes; reduction with hydrogen is also sometimes practicable. Other halogenating agents include ClF_3, AlX_3, CCl_4, hexachloropropene, and BCl_3, while halogen-exchange routes are also employed. Usual actinide starting materials are the metal itself, the oxide or another halide.

To illustrate some of the general points, some routes include:

$$UF_n \xrightarrow[400°]{F_2 \text{ or } ClF_3} UF_6 \qquad (n < 6)$$

$$PaO_2 \xrightarrow[600°]{HF/H_2} PaF_4 \xrightarrow[800°]{F_2} PaF_5$$

$$U_3O_8 \xrightarrow[210°]{\text{hexachloropropene}} UCl_4$$

$$NpO_2 \xrightarrow[500°]{CCl_4} NpCl_4$$

$$NpO_2 \xrightarrow[\text{heat}]{AlBr_3} NpBr_4$$

$$PaCl_5 \xrightarrow[800°]{H_2} PaCl_4$$

$$Th \xrightarrow[400°]{I_2} ThI_4$$

$$PaI_4 \xrightarrow[350°]{\textit{in vacuo}} PaI_3$$

$$UF_6 \xrightarrow[-107°]{BCl_3} UCl_6$$

$$PaBr_5 \xrightarrow[180°]{SiI_4} PaI_5$$

One intriguing question concerns these actinide halides still to be made. AmF_6 has yet to be fully characterised, CmF_6 and EsF_4 are more doubtful, and attempts to make MF_7 (M = Np, Pu) have failed. The short half-lives and intensive radioactivity of the later actinides obviously makes synthesis and study of their compounds difficult, but MX_3 (M = Fm–Lr, X = F, Cl, Br, I) and MX_2 (M = Fm–No) might be expected; possible exceptions would include NoI_3, on account of facile reduction to No^{2+}.

Trihalides

Since they are well-defined for all the actinides except Th and Pa, these are a good vehicle for studying synthetic and structural trends. Allowing for differences arising from redox characteristics, they exhibit a smooth gradation in physical properties across the series.

Fluorides

These all exhibit the LaF_3 structure, with BkF_3 and CfF_3 also obtainable in a YF_3 structure-type. This transition of structural type occurs three places 'later' in the 5*f* series, compared with the 4*f* elements, because of greater radii of the 5*f* metals. Synthetic routes vary; where the +3 state of the actinide is very stable:

$$Ac(OH)_3 \xrightarrow[700°]{HF} AcF_3$$

$$AmO_2 \xrightarrow[heat]{HF} AmF_3$$

$$Cm^{3+}(aq) \xrightarrow{F^-(aq)} CmF_3.xH_2O \xrightarrow[(P_2O_5)]{dry} CmF_3$$

In cases where the 3+ state is slightly (Np, Pu) or very (U) reducing, reductive methods are required:

$$MO_2 \xrightarrow[500°]{HF/H_3} MF_3 \quad (M = Pu, Np)$$

$$UF_4 + Al \xrightarrow{900°} UF_3 + AlF$$

(UH_3 or H_2 are alternative reducing agents)

Chlorides

All the trichlorides made thus far have the 9-coordinate UCl_3 structure: additionally the 8-coordinate $PuBr_3$ type has been characterised for Bk and Cf at high temperatures. It is to be expected that the heavier, smaller, actinides, whose trichlorides have not been isolated, would adopt 6-coordination, probably with the $AlCl_3$ structure.

Two types of preparative route have been employed. A 'conventional' one for actinides forming stable M^{3+} ions, resembles those used for the trichlorides of the 4*f* metals:

$$Ac(OH)_3 \xrightarrow[or\ CCl_4\ 500°]{NH_4Cl\ 250°} AcCl_3$$

$$AmCl_3.6H_2O \xrightarrow[250°]{NH_4Cl} AmCl_3$$

$$M_2O_3 \xrightarrow[400\text{–}600°]{HCl} MCl_3 \quad (M = Es, Cm)$$

Plutonium trichloride can be prepared by various routes:

$$Pu_2(C_2O_4)_3.10H_2O \xrightarrow[or\ C_3Cl_6,\ reflex]{HCl/heat} PuCl_3$$

$$Pu \xrightarrow[450°]{Cl_2} PuCl_3$$

resulting from the non-existence of any higher chlorides.

In the case of uranium and neptunium, reducing methods are used:

$$UCl_4 \xrightarrow[400°]{Al} UCl_3$$

$$UH_3 \xrightarrow[350°]{HCl} UCl_3$$

$$NpCl_4 \xrightarrow[600°]{H_2} NpCl_3$$

$$NpCl_4 \xrightarrow[400°]{NH_3} NpCl_3$$

Bromides

The structures of the tribromides are more complex than those of the fluorides and chlorides. From actinium to neptunium they are 9-coordinate (UCl_3); a second form of $NpBr_3$ as well as the bromides of the following metals up to berkelium exhibit the 8-coordinate $PuBr_3$ structure. This is derived from the tricapped trigonal prismatic UCl_3 structure by removing one of the face-capping halogens (Figure 3.10).

$BkBr_3$ is trimorphic, additionally exhibiting $AlCl_3$ and $FeCl_3$ type 6-coordinate structures in high-temperature modifications. A similar situation probably applies to $CfBr_3$, though here the $PuBr_3$-type structure has only been made indirectly, by β-decay of the corresponding phase of $BkBr_3$. $EsBr_3$ has the $AlCl_3$ structure. Overall, the picture is as expected: that the coordination number decreases as the 5*f* series is traversed.

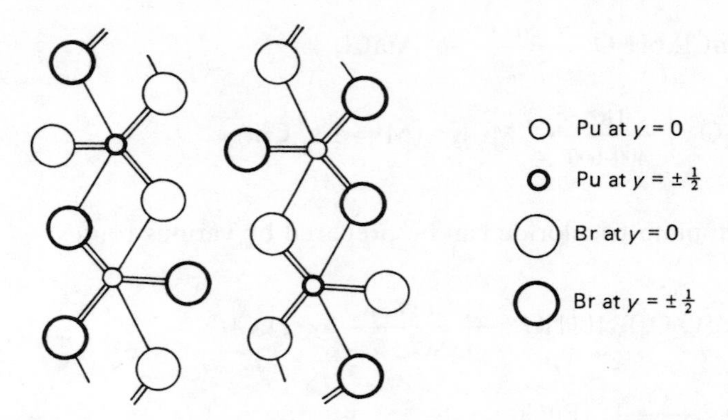

Figure 3.10 The layer structure of $PuBr_3$ [after A. F. Wells, *Structural Inorganic Chemistry*; Clarendon Press, Oxford (3rd edn, 1962)]. Note that 9-coordination is prevented by non bonding Br. . .Br interactions. The notation $y = 0$ and $y = \pm\frac{1}{2}$ indicates the relative height of atoms with respect to the plane of the paper. The planes of the layers are normal to the plane of the paper

Synthetic routes for the bromides include:

$$M_2O_3 \xrightarrow[600\text{–}800°]{HBr} MBr_3 \qquad (M = Cm, Cf)$$

$$Ac_2O_3 \xrightarrow[750°]{AlBr_3} AcBr_3; \qquad U \xrightarrow[500°]{Br_2} UBr_3$$

$$UH_3 \xrightarrow[\text{heat}]{HBr} UBr_3$$

$$NpO_2 \xrightarrow[\text{heat}]{Al/AlBr_3} NpBr_3$$

$$PuBr_3.6H_2O \xrightarrow[70\text{–}120°]{\textit{in vacuo}} PuBr_3$$

Most resemble the chloride routes, though hydrated chlorides cannot be dehydrated by heating. The uranium–bromine route reflects the instability of the higher bromides of uranium. Several methods can usually be applied to the synthesis of one compound, thus $PuBr_3$ can also be made from the reaction of plutonium and bromine, as well as from the reaction of HBr with plutonium(III) oxalate.

Iodides

The trend to decreasing coordination number with increasing size of the halogen is maintained. The incompletely characterised AcI_3 is believed to be 9-coordinate, but the following triiodides from Pa to Am are 8-

coordinate ($PuBr_3$ structure). A high-temperature phase of AmI_3 has the 6-coordinate $FeCl_3$ structure, the type adopted by subsequent actinides.

The synthetic routes below display examples of the general methods established:

$$Ac_2O_3 \xrightarrow[350^\circ]{AlI_3} AcI_3$$

$$MO_2 \xrightarrow[350^\circ]{AlI_3} MI_3 \quad (M = Np, Am)$$

$$PaI_5 \xrightarrow[360^\circ]{\textit{in vacuo}} PaI_3$$

$$UH_3 \xrightarrow[300^\circ]{HI} UI_3$$

$$Pu \xrightarrow[500^\circ]{HgI_2} PuI_3$$

$$CfCl_3 \xrightarrow[540^\circ]{HI} CfI_3$$

$$CmCl_3 \xrightarrow[600^\circ]{NH_4I} CmI_3$$

3.11 Magnetic and spectroscopic properties

The magnetic properties of the actinide ions are more difficult to interpret than those of either the transition metals or the lanthanides. In the latter case, application of the Russell–Saunders coupling scheme leads to a situation where the magnetic properties are those of the ground state, hence the magnetic moment is usually independent of temperature and insensitive to the ion's environment. For the actinides, spin–orbit coupling is much stronger, as are crystal-field splittings, so that the Russell–Saunders scheme is no longer a good approximation – the 'intermediate' coupling (between *RS* and *jj*) scheme is used.

Results for some uranium compounds, as well as some electronic spectra for later actinides, are presented in this section.

3.11.1 Electronic spectra

As with the lanthanides, the *f–f* transitions in the spectra of actinide complexes are relatively weak, though because of the greater size of the more exposed 5*f* orbitals, they interact more with the ligands, leading to higher extinction coefficients and also greater nephelauxetic effects (due to greater covalency). Because of this, absorption bands are shifted from one compound to another to a greater extent than in lanthanide compounds. The electronic dipole transitions are of course 'forbidden', but they are allowed in the presence of an asymmetric ligand field, whether due to a permanent distortion or to temporary coupling with an asymmetric metal–ligand vibration. In addition to the *f–f* transitions, two other kinds of absorption bands are seen in the spectra of actinide complexes. The first type are due to *5f–6d* transitions, and are generally found well above 20 000 cm^{-1} as the *6d* levels are well above the *5f* for most actinides. Thus in the free U^{3+} ion, the $5f^2 6d$ level is more than 30 000 cm^{-1} above the $5f^3$ ground state, while in the solvated U^{3+} aquo-ion the charge-transfer transitions begin around 24 000 cm^{-1}, showing the considerable effect that solvation has on the relative energies of the *5f* and *6d* electrons. The second type are metal–ligand charge-transfer transitions; the maxima are located in the UV region but the tail of an absorption band trailing into the visible region is responsible for the red, brown or yellow colours often noted for actinide complexes with polarisable ligands like Br or I.

Uranium(VI) – UO_2^{2+} – f^0

Although this has a closed-shell electron configuration there is a well-recognised band around 25 000 cm^{-1} (400 nm) with fine structure due to coupling of symmetric O–U–O stretching vibrations to electron-transfer transitions (Figure 3.11).

Uranium(V) – f^1

The ^2F ground state is split into two levels, $^2F_{7/2}$ and $^2F_{5/2}$, in the free ion by spin–orbit coupling; these levels are, however, split under a crystal field; Figure 3.12 shows the effect of an increasing octahedral crystal field on the energy levels, going to the strong field limit. The ground state is a Kramers doublet (ESR signals are seen for U(V) compounds); more importantly, four transitions are expected in the electronic spectrum. In practice, four groups of bands are noted in the electronic spectrum in the range from the near-IR to the visible.

Uranium(IV) – f^2

Published spectra indicate significant difference in both band position and other details between those of 6-coordinate uranium and of higher coordination numbers.

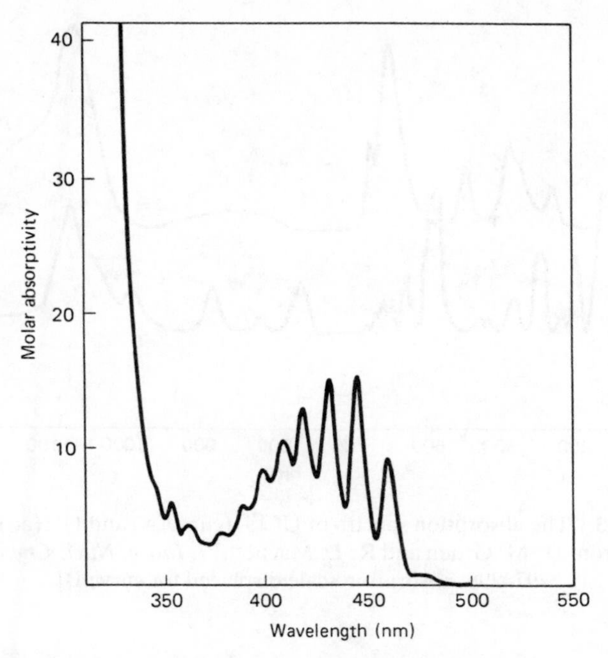

Figure 3.11 Vibronic structure due to O—U—O stretching in the spectrum of $UO_2(O_2CMe)_3^-$ [redrawn from J. L. Ryan and W. E. Keder, *Advan. Chem. Ser.*, **71** (1967) 335]

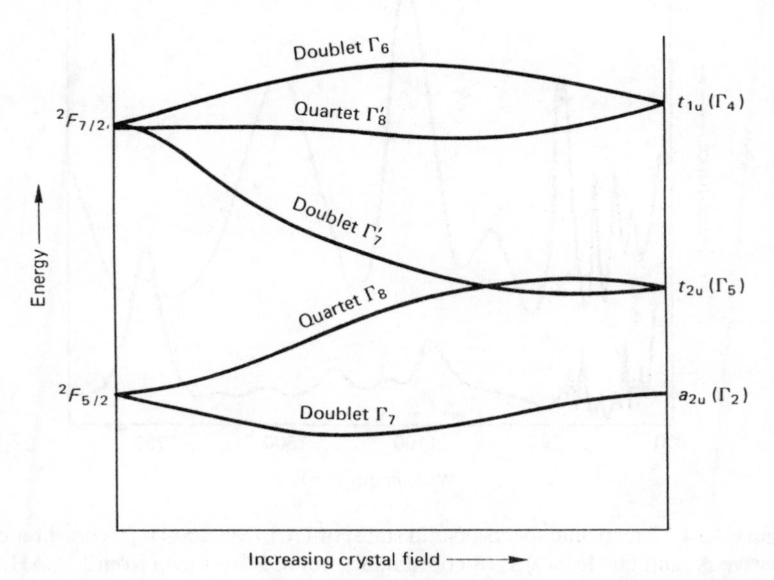

Figure 3.12 The effect of increasing crystal field on the energies of an electron in a f^1 system (uranium(V))

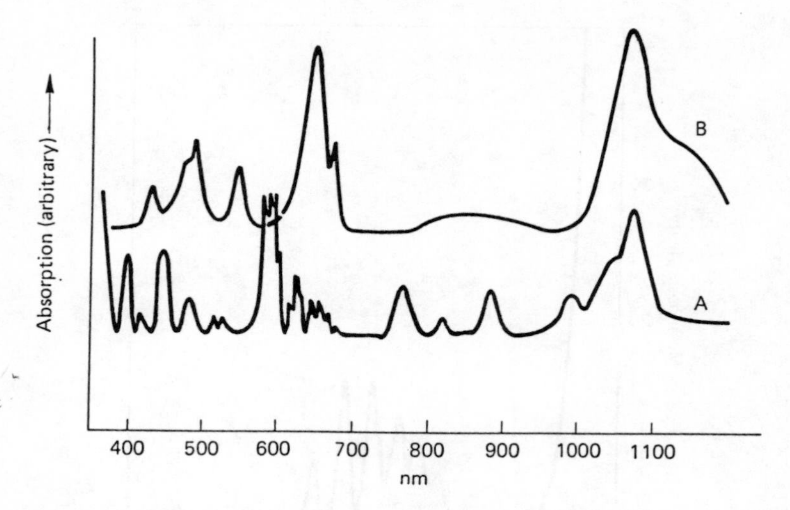

Figure 3.13 The absorption spectra of UCl_6^{2-} (curve A) and $U^{4+}(aq)$ (curve B) [redrawn from D. M. Gruen and R. L. Macbeth, *J. Inorg. Nucl. Chem.*, **9** (1959) 297–298; absorption scale displaced for curve B]

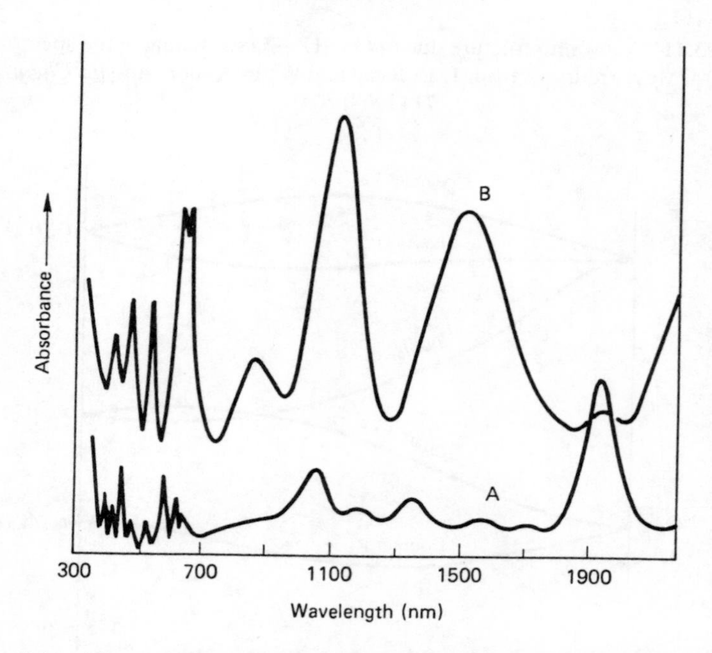

Figure 3.14 Electronic spectra (solid state) of $UCl_4(Me_3CSO)_2$ (6-coordinate), curve A, and $U(Me_2SO)_8I_4$ (8-coordinate), curve B [redrawn from J. G. H. DuPreez and B. Zeelie, *Inorg. Chim. Acta.*, **161** (1989) 187; absorption scale displaced for curve B]

The octahedral UCl_6^{2-} ion has been thoroughly investigated – its spectrum is largely vibronic in nature, with electronic transitions accompanied by vibrations of the complex ion (odd parity modes – the T_{1u} asymmetric stretch and T_{1u} and T_{2u} deformations). The crystal field and spin–orbit couplings are comparable so that there is some overlap of bands from different states. Destroying the centre of symmetry enables pure electronic transitions to be observed (for example, by hydrogen-bonding to UCl_6^{2-}) and alters band patterns in multiplets.

Because of the greater crystal field effects, the $5f$ energy levels are more sensitive to coordination number, and the pronounced difference between UCl_6^{2-} and U^{4+}(aq) (Figure 3.13) has led to the proposal that the aquo-ion does not have a coordination number of 6. Such spectra have been used to support assignment of coordination numbers to uranium(IV) complexes (Figure 3.14).

The later actinides

As an example of the absorption spectrum that can be obtained for these elements, Figure 3.15 shows spectra of a sample of $^{249}BkCl_3$ over a period of time, beginning 11 days after synthesis. ^{249}Bk is a β-emitter with a half-life of 320 d and the spectra obtained over a 976 day (three half-life) period clearly show the loss of the $BkCl_3$ spectrum and its replacement by a spectrum due to $^{249}CfCl_3$. Both spectra are quite reminiscent of the sharp, line-like ones obtained from the lanthanides.

It is of interest that, following the emission of a β-particle from the berkelium nucleus

$$^{249}_{97}Bk^{3+} \longrightarrow {}^{249}_{98}Cf^{4+} + e^-$$

the californium ion regains an electron

$$Cf^{4+} + e^- \longrightarrow Cf^{3+}$$

to maintain the +3 oxidation state. Moreover, the crystal structure type is maintained; X-ray diffraction confirms that the $CfCl_3$ retains the hexagonal structure adopted by the parent $BkCl_3$, rather than changing to the orthorhombic modification.

3.11.2 Magnetic properties

Uranium(VI) compounds should be diamagnetic (f^0, $^1S_0$ ground state); UF_6 and the UO_2^{2+} ion exhibit temperature-independent paramagnetism, explained by coupling of paramagnetic higher-energy states with the ground state.

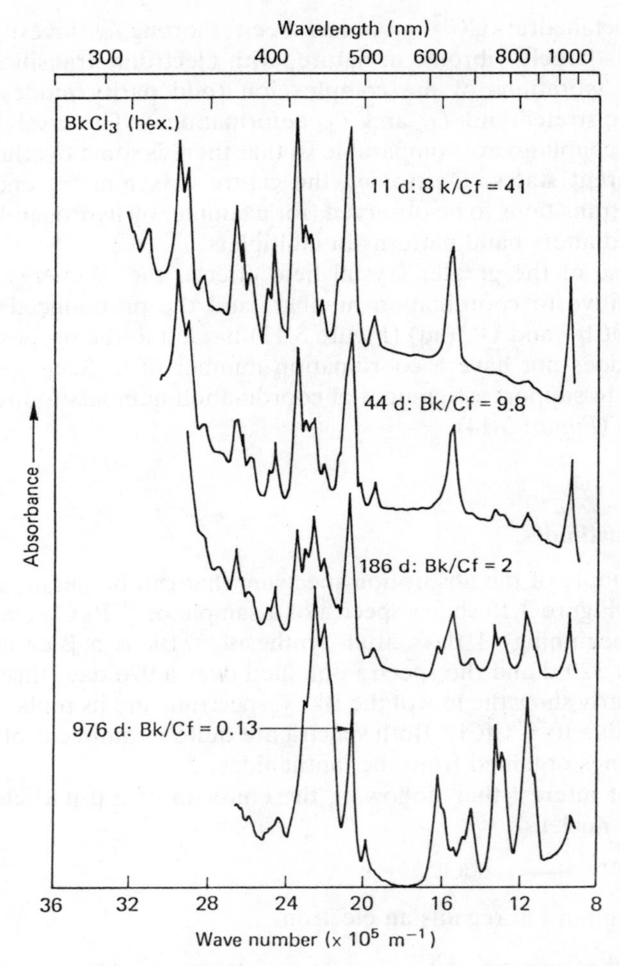

Figure 3.15 The absorption spectrum of the hexagonal form of $BkCl_3$ as a function of time, showing changes due to the appearance of the spectrum of $CfCl_3$ [from J. R. Peterson *et al.*, *Inorg. Chem.*, **25** (1986) 3779; reprinted by permission of the American Chemical Society © 1986]

Uranium(V) compounds are, as expected for a f^1 system, paramagnetic. They generally exhibit Curie–Weiss behaviour, with large Weiss constants.

In the case of uranium(IV), Figure 3.16 shows the successive effects of applying electrostatic interactions and spin–orbit coupling to the f^2 configuration, resulting in a $^3H_4$ ground state. This state undergoes further splitting by a crystal field – the splittings for octahedral and tetragonal geometrics are shown.

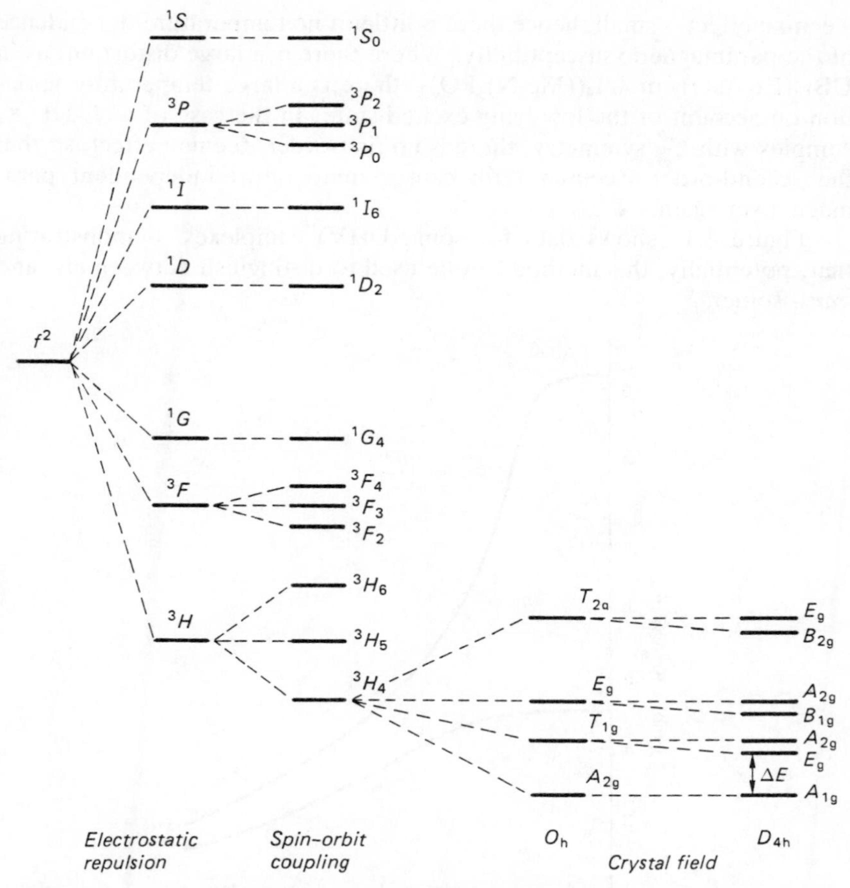

Figure 3.16 A qualitative energy-level diagram showing the successive effects of electrostatic repulsion, spin–orbit coupling and crystal-field splitting on the U^{4+} ion. Crystal-field splittings are only shown for the ground state. The diagram neglects the considerable overlap of levels arising from different states [based, in part, on M. Hirose *et al.*, *Inorg. Chim. Acta*, **150** (1988) L93; reprinted by permission of the Editor]

In a regular octahedral field, there is no contribution to the paramagnetic susceptibility from the first-order Zeeman term. Complexes like UCl_6^{2-} display temperature-independent paramagnetism owing to the second-order Zeeman effect mixing the T_1 excited state into the ground state.

In a *trans*-UL_2X_4 complex of D_{4h} symmetry, both the first- and second-order Zeeman effects contribute to the susceptibility. Where the tetragonal distortion is small (and ΔE in Figure 3.16 is thus big) there is little or no thermal excitation to the excited states, so that the first-order

Zeeman effect is small, hence there is little or no temperature-dependence of the paramagnetic susceptibility. Where there is a large distortion, as in $UBr_4(Et_3AsO)_2$ or $UI_4((Me_2N)_3PO)_2$, there is a large temperature variation on account of the low-lying excited state. In the case of a *cis*-UL_2X_4 complex with C_{2v} symmetry, there is no first-order Zeeman effect, so that the second-order Zeeman term causes temperature-independent paramagnetism again.

Figure 3.17 shows data for some U(IV) complexes, demonstrating that, potentially, this method can be used to distinguish between *cis*- and *trans*-isomers.

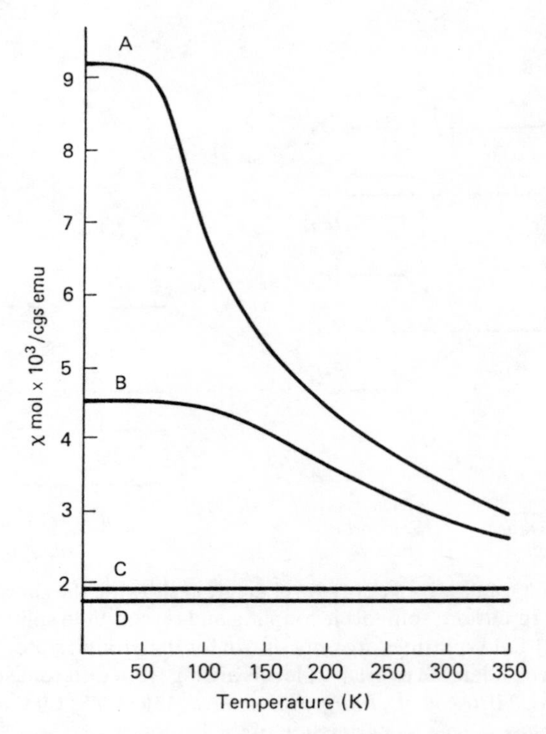

Figure 3.17 Temperature dependence of the magnetic susceptibility of some U(IV) complexes: (A) *trans*-$UBr_4(Et_3AsO)_2$; (B) *trans*-$UCl_4(Et_3AsO)_2$; (C) $(Ph_4P)_2UCl_6$; (D) *cis*-$UCl_4(Ph_3PO)_2$ [redrawn from B. C. Lane and L. M. Venanzi, *Inorg. Chim. Acta*, **3** (1969) 239; reprinted by permission of the Editor]

3.12 Chemistry of the elements – actinium

In its chemistry, actinium strongly resembles the lanthanides in general and lanthanum in particular. It has similar redox potentials (La^{3+}, Ac^{3+} 2.62 V) and ionic radii (La^{3+} 106 pm, Ac^{3+}111 pm). Study of its chemistry is complicated both by its own radioactivity (^{227}Ac = 21.77 y) and by the intense γ-radiation of its own decay products. Much of its chemistry has been investigated on the microgram scale for this reason; in addition, many actinium compounds have been characterised by X-ray diffraction methods, and small-scale work reduces fogging of the film.

Actinium metal can be obtained by reduction of Ac_2O_3, as well as AcF_3 and $AcCl_3$ with alkali metals, as a silvery white solid, sometimes with a golden hue. It forms an fcc lattice with an atomic radius estimated as 198 pm. It is rapidly oxidised to white Ac_2O_3 in moist air.

The halides AcX_3 and oxyhalides AcOX (X = F, Cl, Br) have all been made:

$$Ac(OH)_3 \xrightarrow[300°]{HF} AcF_3 \quad (LaF_3 \text{ structure})$$

$$Ac(OH)_3 \xrightarrow[500°]{CCl_4} AcCl_3 \quad (UCl_3 \text{ structure})$$

$$Ac_2O_3 \xrightarrow[750°]{AlBr_3} AcBr_3 \quad (UCl_3 \text{ structure})$$

$$AcCl_3 \xrightarrow[1000°]{H_2O} AcOCl \quad (PbClF \text{ structure})$$

The similarity with lanthanum is emphasised by the alternative preparation of actinium fluoride by precipitation (At tracer level, actinium coprecipitates qualitatively with lanthanum.)

The oxalate can also be prepared by precipitation; it is isomorphous with $La_2(C_2O_4)_3.10H_2O$ and decomposed on heating:

$$Ac_2(C_2O_4)_3 \longrightarrow Ac_2O_3 + 3CO + 3CO_2$$

The hydrated Ac^{3+} ion can be separated from thorium and the lanthanides by processes such as solvent extraction with thenoyltrifluoroacetone; cation-exchange chromatography is a very efficient way of purifying actinium, as it is much more strongly bound on the resin than its decay products. As expected for a f^0 ion, Ac^{3+} compounds are colourless, with no measurable spectral absorption between 400 and 1000 mm.

3.13 Chemistry of the elements – thorium

In many of the aspects of its chemistry, thorium resembles zirconium and hafnium, exhibiting almost exclusively the +4 oxidation state, rather than its 4*f* analogue, cerium, whose +3 oxidation state is more stable. Thorium chemistry in the +2 and +3 oxidation states is effectively restricted to iodides like ThI_2 and the cyclopentadienyl $Th(C_5H_5)_3$.

Thorium has a wide coordination chemistry, particularly with oxygen-donor ligands, as one would expect for a 'hard' acid.

The metal and its properties

Thorium is a bright, silvery-white metal which tarnishes on exposure to air to a dull black colour. It is soft enough to be scratched with a knife, melts at 1750° and has a density of 11.72 g cm^{-3} (X-ray), similar to lead. Its reactivity is rather similar to magnesium; it dissolves slowly in dilute acid (passivated by HNO_3) with evolution of hydrogen, and can be pyrophoric as a powder.

Typical reactions are shown in Figure 3.18.

The metal can best be obtained by metallothermic reduction of a mixture of ThF_4 with ZnF_2 or $ZnCl_2$, using calcium. The resulting zinc–thorium alloy affords, on distillation of the more volatile zinc, pure thorium. It can also be obtained by the decomposition of ThI_4 on a hot filament.

Binary compounds

Thorium forms a wide range of these compounds. The most important is the oxide ThO_2 (thoria) which has the fluorite structure and melts at 3390°. It gives off a bluish light when heated (with a 1 per cent Ce content, a stronger, whiter, light is obtained) and thus for many years was used in gas mantles.

As already mentioned, direct combination with the elements such as N, C, Si, B, S, P, As and H affords various compounds including Th_3N_4 and ThN, ThC_2, $ThSi_2$, ThB_6, Th_2S_3, Th_3P_4, Th_3As_4 and ThH_2. Th_2S_3 is a high-temperature crucible material, and Th_3P_4 and Th_3As_4 are semiconductors; ThN is a superconductor.

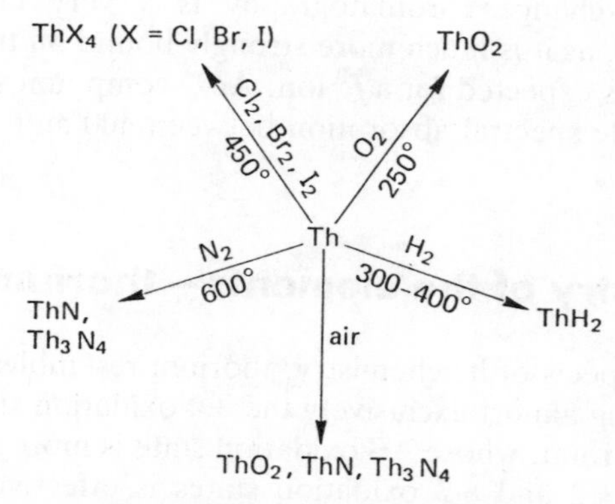

Figure 3.18 Reactions of thorium

Simple salts and aqueous chemistry

The most important salt, crystallising from acid solution as large, colourless, crystals, is $Th(NO_3)_4.5H_2O$ – X-ray diffraction reveals the presence of 11-coordinate thorium, with two uncoordinated water molecules (Figure 3.19).

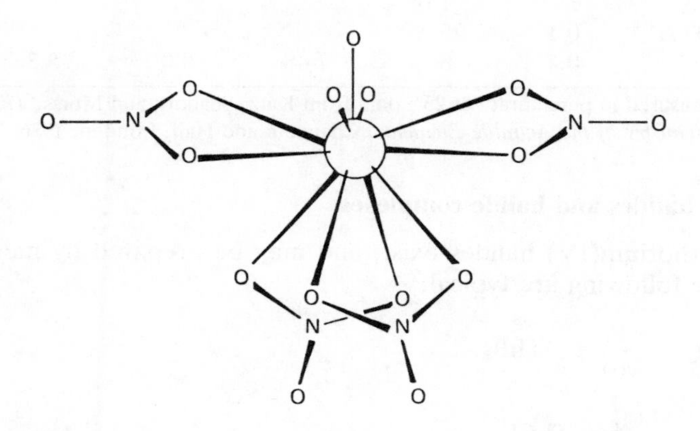

Figure 3.19 The structure of $Th(NO_3)_4.5H_2O$ (hydrogens omitted)

Like uranyl nitrate, it complexes with many O-donor ligands (R_3PO, ROH, R_2O, R_2CO) and can thus be extracted with organic solvents such as tributylphosphate (TBP) doubtless as a complex of the type $Th(NO_3)_4.(TBP)$. Other crystalline salts include $Th(SO_4)_2.xH_2O$ (x = 4, 8, 9) and $Th(ClO_4)_4.4H_2O$. Thorium complexes readily with a wide variety of ligands, several of these complexes being insoluble even in acid solution, such as the phosphate, oxalate, iodate and fluoride. Stability constants for the Th^{4+} complexes of a variety of ligands are listed in Table 3.3; note the preference for fluoride over chloride, and the high *K* values for the chelates like EDTA and, to a lesser extent, acetylacetone. Compared to other 4+ ions, Th^{4+} is relatively unhydrolysed in solution below pH 3. The coordination number in aqueous solution has been assessed at around 9, from low-temperature NMR studies. On hydrolysis, hydroxy-bridged species are formed, possibly involving chains with bridging:

H
O
Th Th
O
H

Table 3.3 Stability constants for Th complexes*

Ligand	I (M)	Log K_1	Log K_2	Log K_3	Log K_4
F^-	0.5	7.56	5.72	4.42	
	4	8.11	6.34		
Cl^-	1	0.18			
SO_4^{2-}	2	3.3	2.42		
NCS^-	1	1.08			
$EDTA^{4-}$	0.1	25.3			
$acac^-$	0.1	8	7.48	6.0	5.3

*Measured in perchlorate at 25°; data from Katz, Seaborg and Morss, *The Chemistry of the Actinide Elements*, Chapman and Hall, London, 1986.

Thorium halides and halide complexes

All four thorium(IV) halides exist, and may be prepared by methods of which the following are typical:

$$ThO_2 \xrightarrow[400^\circ]{F_2} ThF_4$$

$$ThH_2 \xrightarrow{Cl_2} ThCl_4$$

$$Th \xrightarrow[700^\circ]{Br_2} ThBr_4$$

$$Th \xrightarrow[400^\circ]{I_2} ThI_4$$

In all these, thorium is 8-coordinate (see Table 3.4).

Table 3.4 Structures of thorium(IV) halides, ThX_4

	Solid state			Gas phase	
	Coord. no.	Bond lengths (Å)	Approximate polyhedron	Coord. no.	Bond length (Å)
ThF_4	8	2.30–2.37	Square antiprism	4	2.14
$ThCl_4$	8	2.72–2.90 (β) 2.85–2.89 (α)	Dodecahedron	4	2.58
$ThBr_4$	8	2.85–3.12 (β) 2.91–3.02 (α)	Dodecahedron	4	2.72
ThI_4	8	3.13–3.29	Square antiprism		

Many fluoride complexes have been prepared; the formula is often no guide to the coordination polyhedron – thus K_5ThF_9 has isolated ThF_8^{4-} dodecahedra while in M_3ThF_7 (M = Li, NH_4) and $(NH_4)_3ThF_7$, thorium has 9 fluorine neighbours. The $ThCl_6^{2-}$ ion (octahedral) has been characterised in Cs_2ThCl_6 while some bromides and iodides are known, such as $(pyH)_2ThBr_6$ and $(Bu_4N)_2ThI_6$.

Thorium complexes

1. *Oxygen-donor ligands*

As with uranium, a considerable number of these have been isolated for $ThCl_4$, $ThBr_4$, $Th(NCS)_4$ and $Th(NO_3)_4$. Generally the halide complexes are prepared by direct reaction in a weak donor solvent, while the thiocyanates are obtained by metathesis:

$$ThCl_4 + 5Me_2SO \xrightarrow{MeCN} ThCl_4(Me_2SO)_5$$

$$ThCl_4.(Ph_3PO)_2 + 4KNCS \xrightarrow[\text{exc. } Ph_3PO]{Me_2CO} Th(NCS)_4(Ph_3PO)_4 + 4KCl$$

X-ray data are now available for many complexes: the coordination number of thorium is 7 in $ThCl_4L_3$ (L = Ph_3PO, $EtCONMe_2$, $(Me_2N)_3PO$) and 8 in $ThCl_4L_4$ (L = Ph_2SO) and $Th(NCS)_4L_4$ (L = Ph_3PO, $EtCONMe_2$, $(Me_2N)_3PO$). Certain complexes have ionic structures; thus $ThCl_4(Me_2SO)_5$ is $ThCl_3(Me_2SO)_5^+Cl^-$ and $ThCl_4(Me_2SO)_3$ is $ThCl_2(Me_2SO)_6^{2+}ThCl_6^{2-}$ (both, it will be noted, involving 8-coordinate thorium).

Thorium nitrate complexes exhibit a greater variety of structure (see Table 3.5); factors responsible seem to include the small 'bite angle' of nitrate, which increases the feasibility of very high coordination numbers, and the high stability of anionic complexes like $Th(NO_3)_6^{2-}$.

Thorium forms stable complexes with β-diketonates and other chelating ligands; thus $Th(acac)_4$ is a white crystalline solid (mp 171°) which

Table 3.5 Thorium nitrate complexes with O-donor ligands

Formula	Species present	Coordination no.
$MTh(NO_3)_6$ (M = Mg, Ca)	$Th(NO_3)_6^{2-}$	12
$Ph_4PTh(NO_3)_5(Me_3PO)_2$	$Th(NO_3)_5(Me_3PO)_2^-$	12
$Th(NO_3)_4.5H_2O$	$Th(NO_3)_4(H_2O)_3$	11
$Th(NO_3)_4.(Me_3PO)_{2.67}$	$[Th(NO_3)_3(Me_3PO)_4]^+$, $Th(NO_3)_6^{2-}$	10, 12
$Th(NO_3)_4(Ph_3PO)_2$	$Th(NO_3)_4(Ph_3PO)_2$	10
$Th(NO_3)_4(Me_3PO)_5$	$Th(NO_3)_2(Me_3PO)_5^{2+}$	9

sublimes *in vacuo*. It exists in two crystalline forms, both containing 8-coordinate thorium, one dodecahedral, the other square-antiprismatic, showing the small energy difference between these polyhedra; square antiprismatic coordination is also found in $Th(PhCOCHCOPh)_4$. These compounds are generally soluble in covalent solvents and can be utilised in the solvent extraction of trifluorothenoylacetone (TTA)

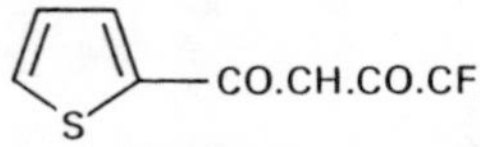

$$-CO.CH.CO.CF_3^-$$

which forms the 8-coordinate $ThTTA_4$. This affords an adduct with trioctylphosphine oxide (TOPO) – $ThTTA_4(TOPO)$ where tricapped trigonal prismatic 9-coordination is found. Similar behaviour is exhibited by the tropolonate (T) adducts ThT_4,L ($L = H_2O$, DMF) where the coordination geometry has been described as a capped square antiprism.

2. *Other donor atoms*

Thiocyanate complexes such as $Th(NCS)_4L_4$ ($L = Ph_3PO$) have already been mentioned; the extremely hygroscopic $Th(NCS)_4(H_2O)_4$ may be similar. A wide range of anionic complexes may be made, usually of the type $M_4Th(NCS)_8$:

$$Th(SO_4)_2 + 2K_2SO_4 + 4Ba(NCS)_2 \xrightarrow{H_2O} 4BaSO_4 + K_4Th(NCS)_8$$

Cubic coordination of Th is found in the tetraethylammonium salt; in the Cs analogue it appears to be antiprismatic. Solid state spectra show two bands in the νC–N region for the former but four bands for the latter; in solution, all have similar two-band spectra, suggesting the anion may be cubic in all of them.

Complexes with neutral donors are less usual but the 7-coordinate $ThCl_4(Me_3N)_3$ has been made by direct reaction.

Alkylamides have been made by conventional salt-elimination reactions. The volatile $Th(NEt_2)_4$ is expected to be dimeric (like the U analogue) but with the bulky silylamide ligand only three chlorines can be replaced:

$$ThCl_4 + 3LiN(SiMe_3)_2 \longrightarrow ClTh[N(SiMe_3)_2]_3 + 3LiCl$$

This is a white crystalline solid, soluble in pentane; it has the structure shown in Figure 3.20, containing 4-coordinate thorium.

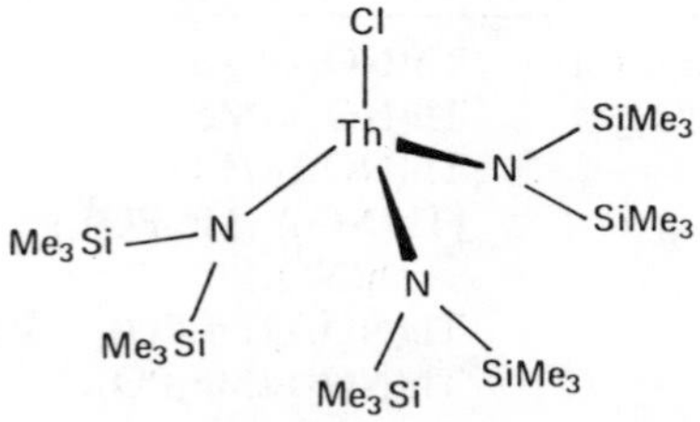

Figure 3.20

The chlorine atom can be replaced with other ligands, such as H, CH_3 or BH_4 to afford similar derivatives $Th(X)[N(SiMe_3)_2]_3$.

Thorium forms few complexes with sulphur as the donor, the best defined being the dithiocarbamates such as the dodecahedral 8-coordinate $Th(S_2CNEt_2)_4$. This can be prepared either by direct reaction of the $ThCl_4$ with NaS_2CNEt_2 or by insertion reactions:

$$Th(NEt_2)_4 + 4CS_2 \longrightarrow Th(S_2CNEt_2)_4$$

Other groups such as CO_2, COS and CSe_2 can be inserted similarly.

A few tertiary phosphine complexes like $ThCl_4(dmpe)_2$ (dmpe = $Me_2P.CH_2.CH_2.PMe_2$) have been made; the chlorines can be replaced to afford other derivatives such as $Th(OPh)_4(dmpe)_2$ and $ThR_4(dmpe)_2$ (R = Me, Ph). All of these are 8-coordinate.

Lower oxidation states of thorium

The only compounds so far reported with an oxidation number other than +4 are certain iodides and the tris(cyclopentadienyl). Heating thorium (IV) iodide with thorium metal affords ThI_2 as well as a compound claimed to be ThI_3, but at present incompletely characterised. ThI_2 has been isolated in α and β forms, the former black and the latter metallic gold in appearance. The observed metallic properties, including a high conductivity, suggest a possible formulation $Th^{4+}(e^-)_2(I^-)_2$.

$Th(C_5H_5)_3$ has been obtained in two forms, too – one form green, the other violet. Both are paramagnetic, as expected for Th(III), and the violet form is reported to be isomorphous with other 4*f* and 5*f* tricyclopentadienyls.

3.14 Chemistry of the elements – protactinium

After uranium and thorium, protactinium has been known longer than any other actinide. The principal isotope, ^{231}Pa, has a half-life of 3.28×10^4 years, making chemical study very feasible, with appropriate radiochemical precautions because of its α-emission. It is formed naturally from ^{235}U:

$$^{235}U \xrightarrow[7.04 \times 10^8 \text{ y}]{\alpha} {}^{231}Th \xrightarrow[25.5 \text{ h}]{\beta^-} {}^{231}Pa$$

Its natural abundance is calculated to be around 0.9×10^{-6} ppm. A considerable amount of protactinium chemistry has been carried out, the major chemical problem being the ready hydrolysis of Pa(V) in solution.

The metal

Protactinium has been prepared by various routes, such as the thermal decomposition of PaI_5 on a tungsten filament or barium reduction of PaF_4 at 1250–1600°.

It is a dense (15.37 g cm^{-3}), malleable, ductile, silvery metal, with a melting point around 1565°. Below 1.4K, it is a superconductor. Some reactions follow:

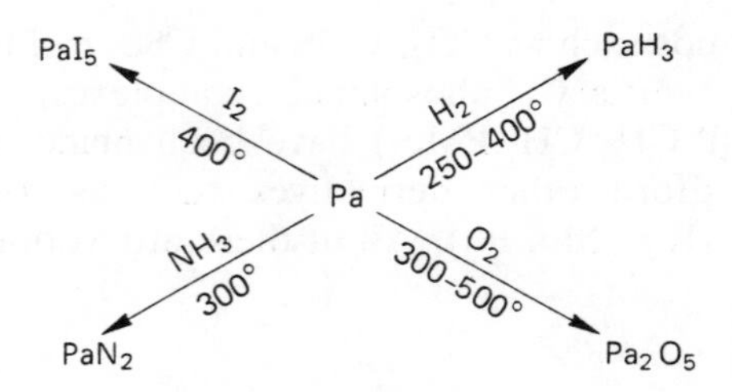

Solution chemistry and the aquo-ions

The solution chemistry of protactinium somewhat resembles niobium and tantalum. The Pa^{5+} ion is hydrolysed in aqueous solution and tends to precipitate as the hydrated oxide $Pa_2O_5.nH_2O$, unless complexing agents like fluoride are present. There is no evidence for an ion PaO_2^+ resembling the uranyl ion. Solutions in fuming nitric acid, however, deposit crystals of $PaO(NO_3)_3.xH_2O$ (x = 1–4) and a similar sulphate complex $PaO(HSO_4)_3$ can be observed.

Air-sensitive solutions of Pa^{4+}(aq) are made by reduction of Pa(V)(aq) with zinc and acid or Cr^{2+} as well as by electrolysis; it only exists in strongly acid solution (such as 3M $HClO_4$). At lower pH, ions like $Pa(OH)_2^{2+}$ or PaO^{2+} and $Pa(OH)_3^+$ are believed to exist.

Halides and oxyhalides

All the halides of Pa(IV) and Pa(V) have been made, as well as PaI_3. Additionally a range of oxyhalides exists, such as PaO_2X (X = F, Cl, Br) and $PaOX_2$ (X = Cl, Br, I).

A variety of preparative routes exists for the halides, commencing usually with Pa_2O_5.

$$Pa_2O_5 \xrightarrow[\text{400–500° } in\ vacuo]{C/Cl_2/CCl_4} \underset{\text{yellow solid}}{PaCl_5} \xrightarrow[\text{or Al/400° } in\ vacuo]{H_2/800°} \underset{\text{(green/yellow solid)}}{PaCl_4}$$

$$Pa_2O_5 \xrightarrow[500°]{HF/H_2} \underset{\text{(brown solid)}}{PaF_4} \xrightarrow[700°]{F_2} \underset{\text{(colourless solid)}}{PaF_5}$$

$$Pa_2O_5 \xrightarrow[600\text{–}700°]{C/Br_2} \underset{\text{(dark red solid)}}{PaBr_5} \xrightarrow[400°\ in\ vacuo]{H_2\text{ or Al}} \underset{\text{(orange–red solid)}}{PaBr_4}$$

$$Pa_2O_5 \xrightarrow[600°\ in\ vacuo]{SiI_4} \underset{\text{(black solid)}}{PaI_5} \xrightarrow[400°,\text{ sealed tube}]{H_2\text{ or Al}} \underset{\text{(dark green solid)}}{PaI_4}$$

$$PaI_5 \xrightarrow{360°\ in\ vacuo} \underset{\text{(dark brown solid)}}{PaI_3}$$

Several structures are known. PaF_5 has the β-UF_5 structure (with 7-coordinate Pa) while $PaCl_5$ has a chain structure (Figure 3.21) also with the unusual 7-coordination (pentagonal bipyramid).

Three crystalline forms of $PaBr_5$ are known; two of these have the α- and β-UCl_4 structures, dimeric with 6-coordinate Pa. PaX_4 (X = F, Cl) have the UF_4 and UCl_4 structures respectively, both with 8-coordination, square antiprismatic (F) and dodecahedral (Cl). $PaBr_4$ also has the UCl_4 structure, while PaI_3 too is 8-coordinate ($PuBr_3$ structure).

Like other early actinides (Th,U) and transition metals (Nb, Ta, Mo, W), protactinium forms several oxyhalides, thus:

$$PaF_5.2H_2O \xrightarrow{250°,\text{ air}} PaO_2F$$

$$PaCl_4 \xrightarrow[150\text{–}200°,\ in\ vacuo]{Sb_2O_3} PaOCl_2$$

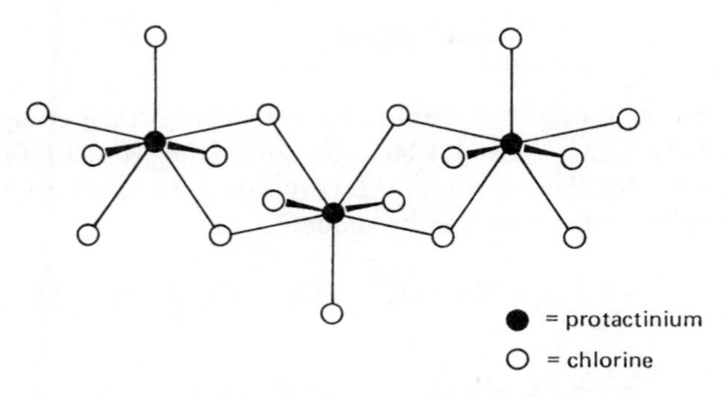

Figure 3.21 The chain structure adopted by $PaCl_5$

$$Pa_2O_5 + PaBr_5 \xrightarrow[\text{sealed tube}]{400^\circ} PaOBr_3$$

$$PaI_5 \xrightarrow[150\text{–}250^\circ,\ \textit{in vacuo}]{Sb_2O_3} PaOI_3,\ PaO_2I$$

While the yellow $PaOBr_3$ has a cross-linked chain structure with 7-coordinated Pa (3O, 4Br), the yellow-green $PaOCl_2$ has a complicated chain structure with three kinds of protactinium, with 7-, 8- and 9-coordination, an important structure which is followed by many other oxyhalides MOX_2 (X = Cl, Br, I; M = Th, Pa, U, Np).

Like many oxyhalides, they tend to decompose or disproportionate on further heating:

$$PaOCl_2 \xrightarrow[\textit{in vacuo}]{550^\circ} PaCl_4 + PaO_2$$

$$PaOBr_3 \xrightarrow[\textit{in vacuo}]{500^\circ} PaO_2Br + PaBr_5$$

$$PaO_2F \xrightarrow{560^\circ} Pa_3O_7F$$

Halide complexes

A large number of these have been synthesised, both anionic complexes and neutral adducts of the halides with donor ligands.

The fluoride complexes of Pa(V) are especially important, both in number and in historical perspective; Von Grosse used K_2PaF_7 in determining the atomic weight of protactinium in 1934. Reaction of HF solutions of Pa(V) and MF (M = K, Rb, Cs) affords $MPaF_6$, M_2PaF_7 and M_3PaF_8 as colourless crystals, the complex obtained depending on the conditions, thus:

$$2KF + PaF_5 \xrightarrow[\text{add acetone}]{17M\ HF} K_2PaF_7$$

X-ray diffraction shows $MPaF_6$ to have dodecahedral 8-coordination of Pa while M_2PaF_7 (M = Cs, K) have 9-coordinate protactinium (capped trigonal prism). Na_3PaF_8 involves the very rare cubic 8-coordination.

Other halide complexes can be made:

$$PaX_4 \xrightarrow[MeCN]{R_4NX} (R_4N)_2PaX_6 \qquad (X = Cl, Br; R = Me, Et)$$

$$PaCl_4 \xrightarrow[SOCl_2/ICl]{CsCl} Cs_2PaCl_6$$

$$PaI_4 \xrightarrow[MeCN]{MePh_3AsI} (MePh_3As)_2PaI_6$$

All probably contain PaX_6^{2-} ions, as has been shown for $(Me_4N)_2PaX_6$ (X = Cl, Br) and Cs_2PaCl_6 (X-ray diffraction).

The complexes with donor ligands can be made by direct reaction of the halides with the ligand in a solution of a suitable polar solvent (acetone, acetonitrile). A range of Pa(IV) complexes, usually yellow in colour, has been made, most having structures similar to the analogous uranium and thorium compounds. Thus $PaX_4[(Me_2N)_3PO]_2$ (*trans*-octahedral), $PaCl_4(Ph_3PO)_2$ (*cis*-octahedral), $PaCl_4(MeCN)_4$ (8-coordinate), $PaCl_4(Me_2SO)_5$ (probably $PaCl_3(Me_2SO)_5^+Cl^-$) and $PaCl_4(Me_2SO)_3$ (possibly $PaCl_2(Me_2SO)_6^{2+}PaCl_6^{2-}$). Some Pa(V) complexes have also been prepared:

$$PaF_5 \xrightarrow[MeCN]{Ph_3PO} PaF_5(Ph_3PO)_2$$

Similar complexes $PaX_5(Ph_3PO)_n$ (X = Cl, Br, n = 1, 2) have been made.

Other complexes

As already noted, protactinium resembles uranium in the complexes it forms. Thus there exists a range of 8-coordinate Pa(IV) β-diketonates like $Pa(PhCOCHCOPh)_4$ and $Pa(CH_3COCHCOCH_3)_4$; the latter has the antiprismatic structure of β-M(acac)$_4$ (M = Th, U, Pu, Np, Pa). These complexes, like the TTA analogue, are soluble in covalent solvents such as benzene and thus suitable for solvent extraction.

The resemblance is similarly found in the thiocyanate $(Et_4N)_4Pa(NCS)_8$ (cubic 8-coordination, as in the thioselenate) and the bis-phthalocyanine. Pa forms a borohydride $Pa(BH_4)_4$ thus:

$$3PaF_4 + 4Al(BH_4)_3 \longrightarrow 3Pa(BH_4)_4 + 4AlF_3$$

Like the uranium analogue, it has a 14-coordinate metal with bridging borohydrides, but its volatility suggests a monomeric gas-phase structure.

The alkoxide $Pa(OEt)_5$ is soluble in benzene but is relatively involatile; it is believed to be hexameric, while the uranium analogue is a dimer.

3.15 Chemistry of the elements – uranium

Numbers of uranium compounds exist in the oxidation states III–VI. Its chemistry, however, is principally that of the +4 and +6 oxidation states; the stability of the oxidation states may be summarised as follows:

U^{3+}	reduces water to hydrogen
U^{4+}	stable in aqueous solution in the absence of air
U^{5+} (UO_2^+)	disproportionates rapidly into a mixture of U(+4) and U(+6) in aqueous solution (slow around pH 2–4)
U^{6+} (UO_2^{2+})	stable in aqueous solution

The relevant redox potentials are:

$$U \xrightarrow{+1.8\ V} U^{3+} \xrightarrow{+0.63\ V} U^{4+} \xrightarrow{-0.62\ V} UO_2^+ \xrightarrow{-0.062\ V} UO_2^{2+}$$

$$U^{4+} \xrightarrow{-0.334\ V} UO_2^{2+}$$

The value for the U/U^{3+} potential suggests that uranium should be a reactive metal (compare Mg/Mg^{2+} 2.38 V, Ti/Ti^{2+} 1.63 V, Mn/Mn^{2+} 1.03 V) and this expectation is borne out in practice as it dissolves rapidly in HCl and HNO_3 (though rather slowly in H_2SO_4 and H_3PO_4).

Uranium metal is very dense (d = 19.0 g cm^{-3}) and chemically very reactive; when pure, it has a silvery appearance, but it is rapidly attacked by air to afford first a yellow film then a black coating of a mixture of oxide and nitride. The powdered metal is usually pyrophoric in air. It reacts readily with hot water and this makes it important to prevent these substances from coming into contact in nuclear reactors. Some of the reactions of uranium are shown in Figure 3.22.

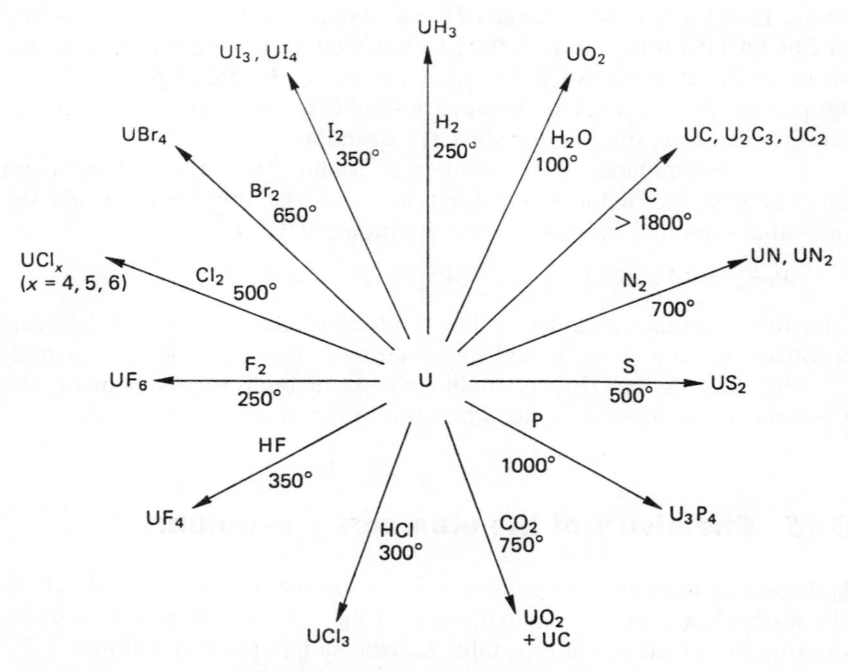

Figure 3.22 Reactions of uranium

Uranium oxides

The story of the uranium oxides is a complex one; some 14 phases have been reported, but not all are genuine, the important ones being UO_2, U_4O_9, U_3O_8 and UO_3.

The brown black UO_2 (fluorite structure) can be prepared thus:

$$UO_3 \xrightarrow[300\text{–}600°]{H_2} UO_2$$

On adding extra oxygen, the resulting UO_{2+x} keeps the fluorite structure until (at a composition that depends on the temperature) the phase U_4O_9 is formed. It has a structure related to the fluorite structure, but with added interstitial oxygens. Several phases have been proposed between U_4O_9 and U_3O_8. The next well-characterised phase, $U_3O_8(UO_{2.67})$, can be conveniently made as a green-black solid by heating uranyl acetate or uranyl nitrate in air:

$$3UO_2(NO_3)_2 \longrightarrow U_3O_8 + 6NO_2 + 2O_2$$

All higher oxides decompose above 650° to U_3O_8, while U_3O_8 loses oxygen above 800°. However, if the high temperature product is cooled in air, it rapidly reverts to U_3O_8, making it a stable weighing form for uranium. U_3O_8 has been described as having a distorted (and oxygen-deficient) UO_3 structure. UO_3 has several crystalline forms; preparative methods include:

$$UO_4.2H_2O \xrightarrow{400°} UO_3$$

$$(NH_4)_2U_2O_7 \xrightarrow{\text{air, } 500°} UO_3$$

$$UO_2(NO_3)_2.6H_2O \xrightarrow[400\text{–}600°]{O_2} UO_3$$

Most forms of UO_3 contain 'uranyl' groups linked by bridging oxygens, but one form (γ) has a structure based on UO_6 octahedra sharing edges and corners.

Uranium(VI)

Apart from a few exceptions – UF_6, UCl_6, UOF_4 and some alkoxides like $U(OMe)_6$ – compounds in this oxidation state are uranyl compounds. Aerial oxidation of any uranium compound eventually results in the formation of a uranyl compound. These can be regarded as derivatives of UO_2^{2+}, the uranyl ion, which is well-characterised from a large number of X-ray diffraction studies, while IR and Raman studies reveal O–U–O stretching frequencies around 860 cm^{-1} (Raman, symmetric) and 920–980 cm^{-1} (IR, asymmetric). Crystallographic study reveals the uranyl ion to be linear, or nearly so, a variety of compounds having U–O distances in the range of about 1.7–1.9 Å. Various relationships have been proposed linking the U–O distance with either the force constant $F_{U\text{-}O}$ or the IR stretching frequency $\nu_{U\text{-}O}$.

The bonding in the uranyl ion (and in the other MO_2^{2+} ions) has been the subject of much debate. It is agreed that the close approach of two oxygen atoms to uranium (U–O ~ 1–7 to 1–9 Å) prevents the close approach of any others (Figure 3.23); *d–p* and *f–p* bonding have been suggested to explain the short U–O bonds.

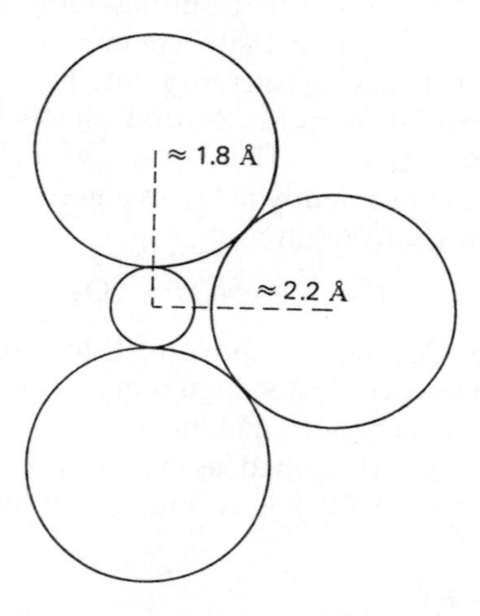

Figure 3.23 The formation of two short U—O bonds in the uranyl ion prevents the close approach of a third oxygen

It is necessary to explain why the UO_2^{2+} ions ($5f^06d^0$) are *trans*-linear, while MoO_2^{2+} ($5d^0$) and WO_2^{2+} ($6d^0$) are *cis* (∠O–M–O ~ 110°), particularly since ThO_2 molecules ($5f^06d^0$) in the gas phase or in matrices are also bent (122°). Most recent work suggests that the difference between UO_2^{2+} and ThO_2 is due to the relative ordering of 5*f* and 6*d*, the former levels being much lower in the uranyl ion. This thus accounts for thorium resembling a *d*-block metal like zirconium. Walsh diagrams suggest that the short U–O bond lengths also favour a *trans*-geometry. (In the MO_2^{2+} (M = Mo, W) ions, where no *f* orbitals are involved, a *cis*-configuration places all the metal 'valence' electrons in bonding MOs and gives maximum use of *d* orbitals in π-bonding.)

Figure 3.24 shows *d* and *f* orbitals of uranium overlapping with oxygen *p* orbitals to give both σ- and π-bonding MOs. A simplified molecular orbital diagram for a uranyl ion is depicted in Figure 3.25. An electron count is as follows: six from uranium, four from each oxygen, deduct two for the 2+ charge, leaving 12 electrons to be accommodated in the low-lying σ_u and σ_g (two each) and π_u and π_g (four each) bonding molecular orbitals.

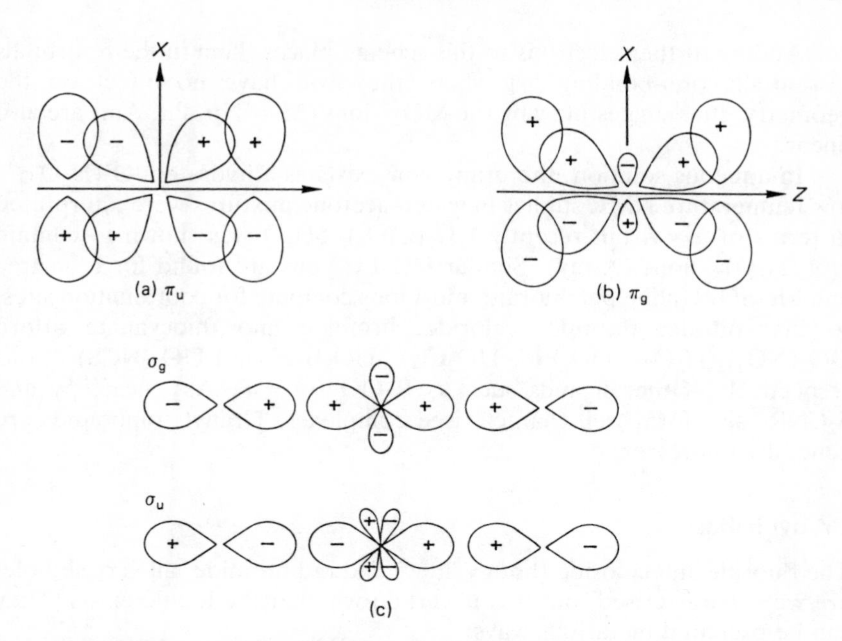

Figure 3.24 π-bonding in the uranyl ion; (a) d_{xz}–p_x; (b) f_{xz^2}–p_x overlaps; (c) σ-bonding in the uranyl ion

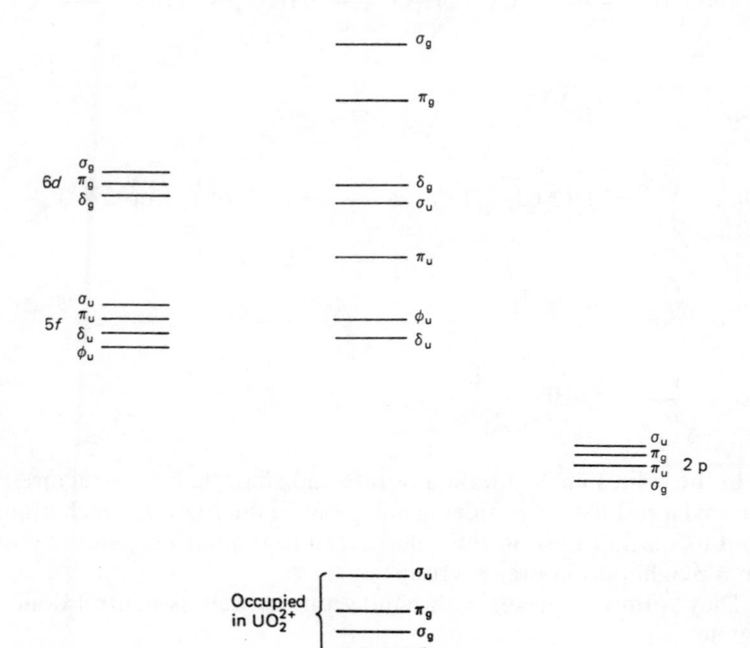

Figure 3.25 MO scheme for UO_2^{2+}

Adding further electrons to this scheme places them in the δ_u orbitals (essentially non-bonding 5*f*) where they will have no effect on the geometry, thus suggesting why the MO_2^{2+} ions (M = Np, Pu, Am) are also linear.

In aqueous solution, the uranyl ion exists as a hydrate $UO_2(H_2O)_n^{2+}$; low temperature NMR studies in water–acetone mixtures were interpreted in terms of $n = 4$ but recently $UO_2(ClO_4)_2.5H_2O$ was shown to contain $UO_2(H_2O)_5^{2+}$ ions (X-ray). Similar $UO_2L_5^{2+}$ ions are found for L = urea and Me_2SO. Unlike perchlorate, most ions compete for coordination sites, so that nitrate, fluoride, chloride, bromide and thiocyanate afford $UO_2(NO_3)_2(H_2O)_2$, $UO_2F_5^{3-}$, $UO_2Cl_4^{2-}$, $UO_2Br_4^{2-}$ and $UO_2(NCS)_5^{3-}$ ions respectively. Other ligands such as R_3QO (Q = P, As), acac, py and S_2CNR_2 also form well-characterised complexes. Uranyl compounds are generally fluorescent.

Uranyl halides

The fluoride and chloride (both yellow) and red bromide, all very soluble, are well-characterised, but it is uncertain whether the iodide exists. They can be prepared in various ways:

$$UO_2(NO_3)_2 \xrightarrow[HNO_3]{H_2O_2} UO_4.2H_2O \xrightarrow{HF} HUO_2F_3.2H_2O \xrightarrow{150^\circ} UO_2F_2$$

$$UO_3 \xrightarrow[400^\circ]{HF} UO_2F_2$$

$$U_3O_8 \xrightarrow[H_2O_2]{HCl} UO_2Cl_2.2H_2O \xrightarrow[450^\circ]{Cl_2/HCl} UO_2Cl_2 \text{ (mp } 578^\circ)$$

$$UCl_4 \xrightarrow[350^\circ]{O_2} UO_2Cl_2$$

$$UBr_4 \xrightarrow[170^\circ]{O_2} UO_2Br_2$$

In the solid state, both the fluoride and chloride have structures based on uranyl ions linked by bridging halogens; in the fluoride, each uranium is bound to six fluorines, in the chloride, to four chlorines, and one oxygen from a neighbouring uranyl group.

They form complexes with halide ions as well as neutral donors, for example:

$UO_2Cl_4^{2-}$, $UO_2Br_4^{2-}$, $UO_2F_5^{3-}$, $UO_2Cl_2(Ph_3PO)_2$

Some uranyl iodide complexes do exist, such as $UO_2I_2(Ph_3AsO)_2$.

Uranyl nitrate

This, the most important uranyl compound, is obtained as yellow crystalline hydrates $UO_2(NO_3)_2.xH_2O$ ($x = 2, 3, 6$) with x varying according to the acidity of the solution (the more concentrated the acid, the lower is x). Its most important property is its high solubility in a very wide range of organic solvents – alcohols, ketones, ethers – as well as in water (quoted solubilities are 0.491, 0.617 and 0.615 g uranyl nitrate per g of ether, acetone and ethanol respectively). This permits the solvent extraction of uranium from aqueous solution, aided by metal nitrates as 'salting-out agents', to make the extraction more effective by increasing the partition coefficient. Tributylphosphate (TBP), $(C_4H_9O)_3PO$, combines the roles of solvent and complexing agent, requiring no salting-out agent, the uranium dissolving as the complex $UO_2(NO_3)_2(TBP)_2$, and this is vital in the separation of uranium and plutonium from fission products in the work-up of spent reactor fuel elements (only MO_2^{2+} form this kind of complex).

Uranyl acetate forms yellow crystals of $UO_2(CH_3COO)_2.2H_2O$ made by dissolving UO_3 in acetic acid. It can be used to precipitate sodium in the presence of magnesium or zinc as the acetate complexes $NaM[(UO_2)(CH_3COO)_3]_3.6H_2O$ containing the $UO_2(CH_3COO)_3^-$ ion. Carbonate complexes are important intermediates in the synthesis of uranium and uranium–plutonium oxide ceramics for use as reactor fuels, and also in the carbonate leaching of uranium ores, which depends on the stability of the complex ion $UO_2(CO_3)_3^{4-}$.

One of the most striking uranyl complexes is the so-called 'uranyl superphthalocyanine', made by heating UO_2Cl_2 with *o*-phthalodinitrile in DMF, as blue-black crystals, subliming at 400° *in vacuo*. The ability of the uranyl ion to sustain 7-coordination is in contrast to other metals, which only afford a tetradentate macrocycle, and in fact 'uranyl superphthalocyanine' undergoes ring-contraction reactions with MX_2 (M = Co, Ni, Cu, Zn etc.) and LnX_3 to give a 'conventional' phthalocyanine (Figure 3.26).

Uranates

Heating U_3O_8 with the oxide, hydroxide or carbonate of an alkali metal affords compounds of the type M_2UO_4 and $M_2U_2O_7$ (M = alkali metal). The group 2 metals form similar compounds as do some transition metals. Under appropriate conditions, other 'uranates' such as M_4UO_5 and M_6UO_6 can be made.

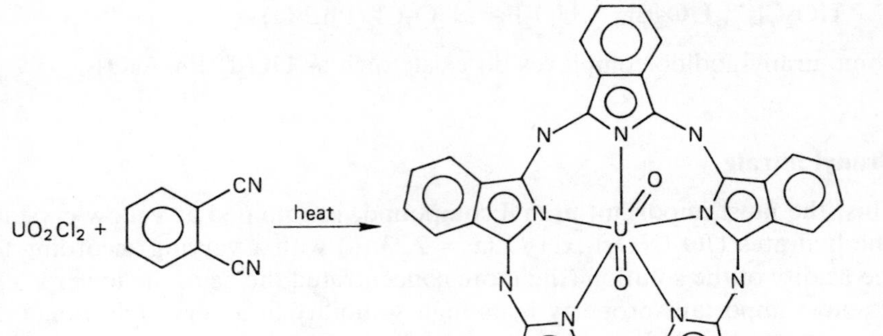

UO_2Cl_2 + [1,2-dicyanobenzene] —heat→ [uranyl superphthalocyanine]

(a)

MX_2 (M = Cu, Zn, Co, for example; X = halide)

(b)

Figure 3.26 (a) The formation of uranyl 'superphthalocyanine' (note that the UO_2 group is perpendicular to the plane of the 5 nitrogens); (b) collapse of the ring to a 'normal' phthalocyanine

The uranates M_2UO_4 do *not* contain UO_4^{2-} ions; instead, they generally have flattened UO_6 octahedra, with two short U–O bonds forming a uranyl group (U–O about 1.9 Å) and four oxygens bridging uraniums. Similar behaviour is followed in $BaUO_4$; on the other hand, in $CaUO_4$ the uranyl group is surrounded by *six* other oxygens. Na_2UO_4 has four short U–O bonds and two longer ones, while there are some compounds like Li_6UO_6 and Cd_2UO_5 with six roughly equal U–O bonds.

Other U(VI) compounds

UOF_4 is another uranium(VI) halide without a uranyl group, prepared by low-temperature hydrolysis of UF_6:

$$UF_6 + H_2O \longrightarrow UOF_4 + 2HF$$

The short U–O bond (1.78 Å) does, again, argue for U–O multiple bond character.

The hexahalides UF_6 and UCl_6 are both octahedral molecules. The colourless fluoride is especially important as its volatility (sublimes at 56.5°; mp 64° under pressure) is responsible for its use (on the ton scale) in isotope separation (section 3.6). Many methods exist for its synthesis, for example

$$UF_4 \xrightarrow[250\text{–}400°]{F_2} UF_6$$

$$2UF_4 \xrightarrow[600\text{–}900°]{O_2} UO_2F_2 + UF_6$$

the latter, the Fluorox method, avoiding the use of elemental fluorine.

UCl_6 is a thermally sensitive and hygroscopic dark green solid, mp 177.5° (but volatile below 100° *in vacuo*):

$$2UCl_5 \xrightarrow{120°} UCl_4 + UCl_6$$

$$UF_6 \xrightarrow[-107°]{BCl_3} UCl_6$$

The alkoxides are discussed under uranium(V)

A very wide range of uranyl complexes has been synthesised and many of these have been studied by X-ray diffraction. The results show that the uranium ion is surrounded by a 'girdle' of 4, 5 or 6 ligands which are coplanar (or nearly so) with a plane normal to the O–U–O axis; in general, the atoms are coplanar if there are 4 or 5 of them but form a puckered ring if there are 6, unless they are bidentate ligands with a small 'bite angle' (such as nitrate).

Table 3.6 Representative uranyl complexes

6-coordinate (2 + 4)	7-coordinate (2 + 5)	8-coordinate (2 + 6)
$MgUO_4$	$UO_2L_5^{2+}$ (L = H_2O, urea, DMSO)	UO_2F_2
$BaUO_4$	UO_2Cl_2	UO_2CO_3
$Cs_2UO_2Cl_4$	$UO_2F_5^{3-}$	$UO_2(CO_3)_3^{4-}$
$(Me_4N)_2UO_2Br_4$	$UO_2(NCS)_5^{3-}$	$UO_2(NO_3)_2.2H_2O$
	$UO_2(OAc)_2(Ph_3RO)$ (R = P, As)	$CaUO_4$
	$UO_2(hfac)_2.L$ (L = THF, NH_3, Me_3PO)	$SrUO_4$
	UO_2superphthalocyanine	$RbUO_2(NO_3)_3$

Uranium(V)

It is firmly believed that the UO_2^+ ion exists, but no compound containing it (except possibly UO_2Cl and UO_2Br) has yet been isolated in the solid state. However, the one-electron polarographic reduction of UO_2^{2+} to UO_2^+ is reversible (unlike the second step, to U^{4+}) which argues for the survival of the UO_2 moiety. Furthermore, other actinides (Np, Pu, Am) do form stable MO_2^+ ions, isolable in crystalline compounds.

Part of the problem with characterising UO_2^+ is that, in aqueous solution, it has only a limited stability range, around pH 2, when the disproportionation $2UO_2^+ \rightarrow UO_2^{2+} + U^{4+}$ is relatively slow. It is stabler in certain non-aqueous solvents such as Me_2SO, while U(V) can be stabilised in halides and halo complexes like UX_5, UX_6^- and UOX_5^{2-}.

The halides (except UI_5) have been made by routes such as:

$$UF_6 \xrightarrow{HBr} UF_5 \text{ (grey/pale yellow solid)}$$

$$UO_3 \xrightarrow[160°,\ 20\ atm]{CCl_4} UCl_5 \text{ (red-brown solid)}$$

$$U \xrightarrow[MeCN\ Cat.]{Br_2} UBr_5 \text{ (brown solid)}$$

$$UBr_4 \xrightarrow[soxhlet]{Br_2} UBr_5$$

They tend to be unstable thus:

$$2UCl_5 \xrightarrow{120°} UCl_4 + UCl_6$$

UF_5 has a polymeric structure with 6- and 7-coordinate uranium; UCl_5 is dimeric while one form of UBr_5 has the same structure.

UF_5 dissolves in HF(aq) to give a stable U(V) solution from which salts like the blue $CsUF_6$ may be crystallised; others may be made thus:

$$2UO_3 + 6SOX_2 + 2MX \longrightarrow 2MUX_6 + SO_2X_2 + 5SO_2$$

$$(M = Rb, Cs;\ X = Cl, Br)$$

as stable crystalline salts: the chlorides are yellow and the bromides brown to black.

Oxyhalides UO_2X and UOX_3 (X = Cl, Br) have been made by several methods:

$$UO_2Cl_2 + UO_2 \xrightarrow{590°} 2UO_2Cl$$

$$UO_3 \xrightarrow[250°]{HBr/N_2} UO_2Br$$

$$UO_2Br_2 \xrightarrow{\text{heat}} UO_2Br$$

$$UCl_4 + UO_2Cl_2 \xrightarrow{370°} 2UOCl_3$$

$$UO_3 \xrightarrow[110°]{CBr_4/N_2} UOBr$$

Of these, only UO_2Br has had its structure determined.

The oxyhaloanions can conveniently be made by oxygen-abstruction:

$$UX_6^- \xrightarrow{Me_2CO} UOX_5^{2-}$$

and isolated as pastel-coloured Et_4N^+ salts (F, pink; Cl, blue; Br, green).

In addition to these, a number of halide adducts have been made, such as $UF_5(Ph_3PO)_n$ (n = 1, 2) and $UX_5(Ph_3PO)$ (X = Cl, Br), as well as adducts with ligands such as pyridine and phenanthroline, and the synthetically useful $UCl_5.SOCl_2$. Few structural data are available.

The alkoxides are the most studied U(V) compounds and in fact a wide variety have been made in the oxidation states (IV)–(VI). Thus

$$UCl_5.SOCl_2 \xrightarrow[EtOH]{py} (pyH)_2UOCl_5 \xrightarrow[NH_3]{EtOH} UO(OEt)_3 \longrightarrow U(OEt)_5$$

$$U(OEt)_5 + 5ROH \longrightarrow U(OR)_5 + 5EtOH$$

$U(OEt)_5$ is a dark-brown liquid, stable to 170° and distilling at 123° (0.001 mm pressure); it is associated (dimer?) in solution but $U(OCEt_3)_5$ seems to be a monomer.

The U(V) alkoxides can be oxidised to the very volatile (and moisture-sensitive) $U(OR)_6$, which are probably all octahedral:

$$U(OEt)_5 \xrightarrow[\text{peroxide}]{\text{benzoyl}} U(OEt)_6$$

$U(OMe)_6$, which distils at 87° (0.01 mm), has attracted attention for its possible utility in the laser isotope separation of ^{235}U (section 3.6); $U(OCH_2CF_3)_6$ is exceptionally volatile (bp 25° at 10 mm).

The green uranium(IV) alkoxides, $U(OR)_4$, tend to be non-volatile, highly-associated solids, conveniently prepared either by salt-elimination reactions or by alcoholysis of the diethylamide:

$$UCl_4 + 4NaOR \longrightarrow U(OR)_4 + 4NaCl$$

$$U(NEt_2)_4 + 4ROH \longrightarrow U(OR)_4 + 4Et_2NH$$

With bulky ligands, monomeric structures become possible, as in the orange tetrakis(2,6-di-t-butylphenoxy) uranium (Figure 3.27).

In addition to the foregoing, other alkoxides are known with U_3O cores and mixed valencies, like $U_2(OBu^t)_9$.

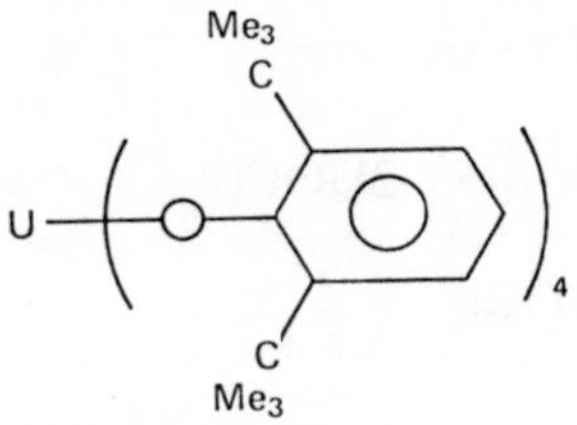

Figure 3.27

Uranium(IV)

U^{4+} ions are stable in aqueous solution in the absence of air, largely as the apple-green $U^{4+}(aq)$ ion; it can be made by reduction of UO_2^{2+} with zinc and sulphuric acid. There is some hydrolysis, reduced by acidification. A variety of U(IV) salts can be obtained, most, like the perchlorate $U(ClO_4)_4.xH_2O$ by crystallisation, but some like the fluoride and iodate by precipitation.

The most important binary compounds are the oxide UO_2 and the halides UX_4 (X = F, Cl, Br, I). The brown-black UO_2 is part of the complicated uranium oxide story (see page 127); it has the fluorite structure with 8-coordinate uranium, and is made by:

$$U_3O_8 \text{ or } UO_3 \xrightarrow[300\text{–}600^\circ]{CO \text{ or } H_2} UO_2$$

It tends to re-oxidise to U_3O_8 in air.

The tetrahalides may be prepared by a variety of methods:

$$UO_2 \xrightarrow[\text{heat}]{HF(g)} UF_4$$

$$U_3O_8 \xrightarrow[\text{reflux}]{C_3Cl_6} UCl_4$$

$$U \xrightarrow[650^\circ]{Br_2/He} UBr_4$$

$$U \xrightarrow[500^\circ/20 \text{ kPa}]{I_2} UI_4$$

They have a variety of structures, both the fluorides (square antiprism) and chloride (dodecahedral) having 8-coordination. UBr_4 has a bridged structure with 7-coordinate (pentagonal bipyramid) uranium while UI_4 has a zig-zag chain structure based on UI_6 octahedra sharing edges.

The trend of decreasing coordination number with increasing size of the halogen is to be expected.

Table 3.7 The uranium(IV) halides

	UF_4	UCl_4	UBr_4	UI_4
Description	Air-stable green solid	Deliquescent green solid	Deliquescent brown solid	Deliquescent black solid
mp°	1036	590	519	506
bp°		789	761	
Solubility	—	(soluble in H_2O and common organic solvents)		
Coordination geometry	Square antiprism	Dodecahedron	Pentagonal bipyramid	Octahedral
CN of metal	8	8	7	6
U–X (Å)	2.25–2.32	2.64–2.87	2.61 (terminal) 2.78–2.95 (bridge)	2.92 (terminal) 3.08–3.11 (bridge)

Apart from the insoluble UF_4, they are soluble in covalent solvents (such as Me_2CO, MeCN); these solutions are useful for preparing U(IV) complexes.

Many anionic halide complexes have been synthesised. Some have been made by precipitation with big cations (Cs^+, Ph_4P^+, Me_4N^+) from concentrated acid solution; others by mixing the constituents in a non-polar solvent, thus:

$$2CsCl + UCl_4 \xrightarrow{HCl(aq)} Cs_2UCl_6$$

$$2Ph_4AsI + UI_4 \xrightarrow{MeCN} (Ph_4As)UI_6$$

$$UF_4 + 4NH_4F \xrightarrow{100^\circ} (NH_4)_4UF_8$$

The complexes are generally green.

A number of X-ray studies have been undertaken: in the fluoride complexes, uranium commonly has a coordination number of 8 or 9; in the others, 6. Some of the fluoro complexes contain discrete ions, like UF_8^{4-} (distorted tetragonal antiprism) in $(NH_4)_4UF_8$.

K_3UF_7 has pentagonal-bipyramidal 7-coordination. 9-coordination is found in K_2UF_6 (trigonal prismatic) and dodecahedral 8-coordination in Rb_2UF_6 and $CaUF_6$, the high coordination numbers being attained by fluoride bridging.

Apart from these compounds, uranium(IV) complexes can mostly be assigned to one of two classes; (i) adducts of halides (including nitrates and pseudo-halides) and (ii) chelate compounds.

(i) *Adducts of halides*

Over the last 20 years or so, a considerable number of these complexes have been isolated. Most of these have involved ligands which bind to uranium through an oxygen atom, such as tertiary phosphine and arsine oxides, sulphoxides and amides, typical examples being Ph_3PO, Et_3AsO, Me_2SO, $(Me_2N)_3PO$ (hexamethylphosphoramide) and $EtCONEt_2$. They are generally prepared by direct reaction between the halide and ligand in a non-aqueous solvent such as acetone, acetonitrile, nitromethane, or tetrahydrofuran, for example

$$UBr_4 + 2Ph_3PO \xrightarrow{MeNO_2} UBr_4(Ph_3PO)_2$$

Nitrate complexes are best prepared from $Cs_2U(NO_3)_6$ while thiocyanate complexes are obtained by metathesis of the halide complexes with KNCS in acetone. After the precipitates of $CsNO_3$ and KX (X = Cl, Br) have been filtered off, the complexes can be crystallised on concentrating the filtrate:

$$Cs_2U(NO_3)_6 + 2Ph_3AsO \xrightarrow{Me_2CO} U(NO_3)_4(Ph_3AsO)_2 + 2CsNO_3$$

$$UCl_4 + 4KNCS + 4(Me_2N)_3PO \xrightarrow{MeNO_2} U(NCS)_4[(Me_2N)_3PO]_4 + 4KCl$$

The stoichiometry of the complexes varies from $UX_4.2L$ to $UX_4.7L$; sometimes more than one complex can be obtained, depending on the choice of solvent and quantity of ligand – thus $UCl_4(Me_2SO)_n$ (n = 3, 5, 7). Coordination of the ligands via oxygen has been indicated by IR spectra – thus there is a shift in $\nu(S{=}O)$ from 1050 cm^{-1} (free ligand) to 960 cm^{-1} in $UCl_4(MeSO)_3$. Inclusion of dimethyl sulphoxide of crystallisation (probably 2 molecules) was deduced from the observation of two bands (at 1047 cm^{-1} and 942 cm^{-1}) in the IR spectrum of $UCl_4(Me_2SO)_7$.

Extensive X-ray diffraction studies have confirmed the structures of numerous complexes: UCl_4L_2 (L = Et_3AsO, $(Me_2N)_3PO$, $(Me_2N)_2PhPO$), UBr_4L_2 (L = Ph_3PO, Ph_3AsO, $(Me_2N)_3PO$), these being *trans*-octahedral; $U(NCS)_4L_4$ (L = Ph_3PO, Me_3PO, $(Me_2N)_3PO$), all square-antiprismatic; and $U(NO_3)_4(Ph_3PO)_2$ (*trans* – with bidentate nitrates).

Some however, are more surprising: $UCl_4(Me_3PO)_6$ is $UCl(Me_3PO)_6^{3+}(Cl^-)_3$ while $UCl_4(Me_2SO)_5$ is $UCl_3(Me_2SO)_5^+Cl^-$ and $UCl_4(Me_2SO)_3$ is $UCl_2(Me_2SO)_6^{2+}UCl_6^{2-}$. The 2.5:1 complexes $UCl_4(RCONR_2)_{2.5}$ (R = Me, Et) are $UCl_3(RCONR_2)_4^+UCl_5(RCONR_2)^-$, and $UI_4(Ph_3AsO)_2$ is $UI_2(Ph_3AsO)_4^{2+}UI_6^{2-}$.

Similar adducts have been isolated with other donor atoms. With nitrogen, examples are $UCl_4(Me_3N)_2$, which has a *trans*-octahedral structure (X-ray), and UCl_4py_2. Many early attempts to prepare phosphine complexes were unsuccessful, owing to oxidation of the ligand, but well-characterised examples include $UCl_4(PMe_3)_3$ and $UCl_4(Me_2P(CH_2)_2PMe_2)$.

One reason for the greater ease in obtaining adducts with oxygen donor ligands is the dipolar contribution of Ph_3PO, though no complexes of similar monodentate sulphur ligands have been obtained (but note the 8-coordinate dithiocarbamates such as $U(S_2CNEt_2)_4$).

Two case studies

(a) *The UX_4–Ph_3PO complexes*

When a solution of triphenylphosphine oxide in a solvent such as tetrahydrofuran is added to a similar solution of UCl_4, a pale green precipitate analysing as $UCl_4(Ph_3PO)_2$ forms. X-ray diffraction studies published in 1975 on crystals obtained by recrystallisation from nitromethane revealed them to contain *cis*-$UCl_4(Ph_3PO)_2$; the phenyl rings in neighbouring Ph_3PO molecules face each other with an interplanar distance of some 3.52 Å. The authors suggested that this graphite-type interaction was consistent with some interaction between the rings and that "the *cis*- was preferred to the *trans*-configuration in order to achieve more stable crystal packing." (All UX_4L_2 systems subsequently studied by X-ray diffraction have been the *trans*-isomers.)

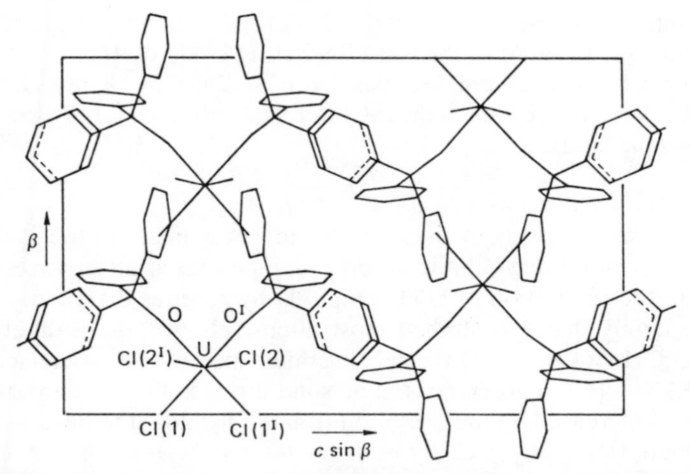

Figure 3.28 The structure of *cis*-$UCl_4(Ph_3PO)_2$ showing the graphite-like packing of certain phenyl groups, indicating that π-interaction in the *cis–isomer* may favour its isolation [from G. Bombieri *et al.*, *J. Chem. Soc. Dalton*, (1975) 1875]

Subsequently it was noticed that the IR spectrum of $UCl_4(Ph_3PO)_2$ depended on its history. Crystals isolated immediately after precipitation had a different spectrum from that obtained from authentic samples of the *cis*-isomer, though it did resemble the spectrum of *trans*-$UBr_4(Ph_3PO)_2$. Further study suggested, on the basis of electronic spectra and X-ray powder photographs, that the 'anomalous' crystals were the *trans*-isomer. It seems that the less soluble *trans*-form crystallises first, but on recrystallisation – or merely being left in contact with solvent – it rearranges to the *cis*-form.

(b) *Uranium(IV) nitrate complexes of hexamethylphosphoramide*
Reaction of $Cs_2U(NO_3)_6$ with 2 moles of hexamethylphosphoramide affords a dark green crystalline complex $U(NO_3)_4[(Me_2N)_3PO]_2$; the IR spectrum shows none of the characteristic bands due to ionic nitrate and it seems probable that this is a 10-coordinate complex with bidentate nitrate coordination, like $U(NO_3)_4[Ph_3PO]_2$.

Reaction of $U(NO_3)_4[(Me_2N)_3PO]_2$ with excess ligand affords light green crystals of $U(NO_3)_4[(Me_2N)_3PO]_4$. This has a more complex IR spectrum, apparently corresponding to two types of nitrate coordination, but on dissolving in solvents such as MeCN or $MeNO_2$, it simplifies, possibly owing to loss of ligand to afford an 11-coordinate complex of the type $U(NO_3)_4[(Me_2N)_3PO]_3$. The use of large anions (ClO_4^-, PF_6^-, BF_4^-, BPh_4^-) to 'fish out' cations from solution is well-known to chemists, and crystals of $U(NO_3)_3[(Me_2N)_3PO]_3^+BPh_4^-$ were obtained from solutions of $U(NO_3)_4[(Me_2N)_3PO]_2$ and excess ligand by treatment with $NaBPh_4$. The IR spectrum of this complex shows only one mode of nitrate coordination, significantly different from $U(NO_3)_4[(NMe_2)_3PO]_4$ and suggesting that 9-coordination is found in the ion $U(NO_3)_3[(Me_2N)_3PO]_3^+$.

None of these complexes has been studied by X-ray diffraction methods, but the results indicate inferences that can be drawn from spectroscopic study.

(ii) *Chelate compounds*
The first intensive study of these compounds was made in the 1940s with the objective of finding volatile uranium compounds as alternatives to UF_6 for use in the separation of ^{235}U from ^{238}U by gaseous diffusion.

The family that was studied most intensively was the β-diketonates, $U(R_1CO.CH.CO.R_2)_4$, a typical example being the acetylacetonate ($R_1 = R_2 = CH_3$), a green crystalline solid mp 175–176°; even though the volatility is increased by the use of fluorinated ligands (the hexafluoroacetylacetonate ($R_1 = R_2 = CF_3$) has mp = 60° and boils at 70° at 0.001 mm pressure), none were volatile enough to supplant UF_6. The complex with 2-thenoyltrifluoroacetone (TTA)

S —CO—CH—CO—CF_3

and its plutonium analogue are involved in one of the processes for separating uranium and plutonium.

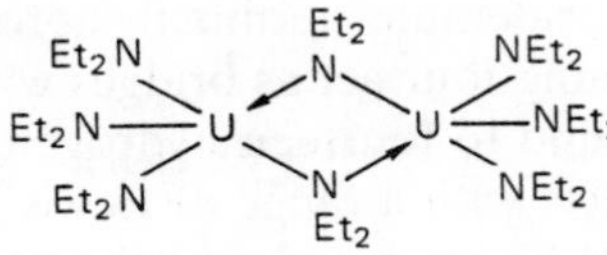

Figure 3.29

A number of uranium(IV) amides have been characterised. Uranium diethylamide (but not other alkylamides) can be prepared by a salt elimination reaction:

$$UCl_4 + 4LiNEt_2 \longrightarrow U(NEt_2)_4 + 4LiCl$$

It is a green solid (mp 36°) which can be distilled at 80° *in vacuo*. In the solid state it has a dimeric structure (Figure 3.29).

Using the bulkier diphenylamide gives the red monomeric $U(NPh_2)_4$ (tetrahedral) while the even bulkier bis(trimethylsilyl)amide can only replace three chlorines to afford $UCl[N(SiMe_3)_2]_3$, doubtless 4-coordinate, with a chemistry quite analogous to that of the corresponding thorium compound (see section 3.13).

Uranium borohydride can be prepared thus:

$$UF_4 + 2Al(BH_4)_3 \longrightarrow U(BH_4)_4 + 2AlF_2(BH_4)$$

It forms volatile dark green crystals.

It doubtless forms monomeric molecules in the vapour phase, like the zirconium and hafnium analogues, but in the solid state has a complex polymeric structure (Figure 3.30) in which each uranium has six boron 'neighbours'.

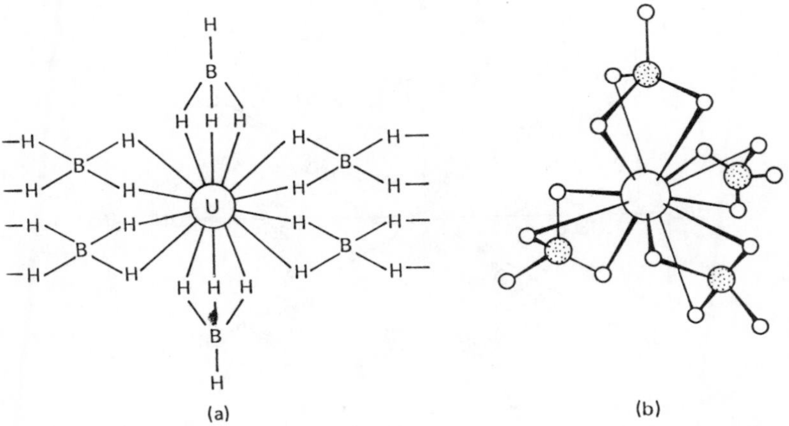

(a) (b)

Figure 3.30 Structures of $U(BH_4)_4$: (a) solid-state 'polymer'; (b) possible vapour-phase monomer

Two of these are tridentate terminal borohydrides, with UH_3B bridges, while the remaining four act as bridges with UH_2BH_2U bonding. Each uranium thus is bound to fourteen hydrogen atoms.

$U(BH_4)_4$ forms adducts with a range of Lewis bases, with a variety of structures. $U(BH_4)_4(THF)_2$ is pseudo-octahedral with, formally, 14-coordinate U (all BH_4 groups tridentate), but $U(BH_4)_4(Et_2O)$ has a polymeric structure and $U(BH_4)_4[(C_3H_7)_2O]$ is an asymmetric trimer.

Uranium(III)

U^{3+} ions may be generated in solution simply by dissolving UCl_3 in water or by the reduction of UO_2^{2+}, either by chemical reduction with zinc amalgam or by electrolytic reduction using a mercury cathode and platinum anode. If the latter course is followed using uranyl sulphate, ethanol is added to the resulting solution and red crystals of $U_2(SO_4)_3.8H_2O$ are formed. Hydrated UF_3 may be obtained similarly, while addition of concentrated HCl to the sulphate solution affords ultimately halide complexes like $MUCl_4.5H_2O$ (M = K, Rb, NH_4). If, however, the U^{3+} solution is kept, it oxidises by attacking the water – more rapidly if air is present:

$$U^{3+}(aq) + H_2O(l) \longrightarrow U^{4+}(aq) + \tfrac{1}{2}H_2(g) + OH^-(aq)$$

Uranium metal is attacked by hydrogen at a temperature of 250° (or less, if the metal is powdered); as attack proceeds, the product swells up and changes into a fine black powder, UH_3. Although it is very reactive, being pyrophoric in air, it is the most useful uranium(III) compound. Some of the reactions of uranium(III) hydride are given below.

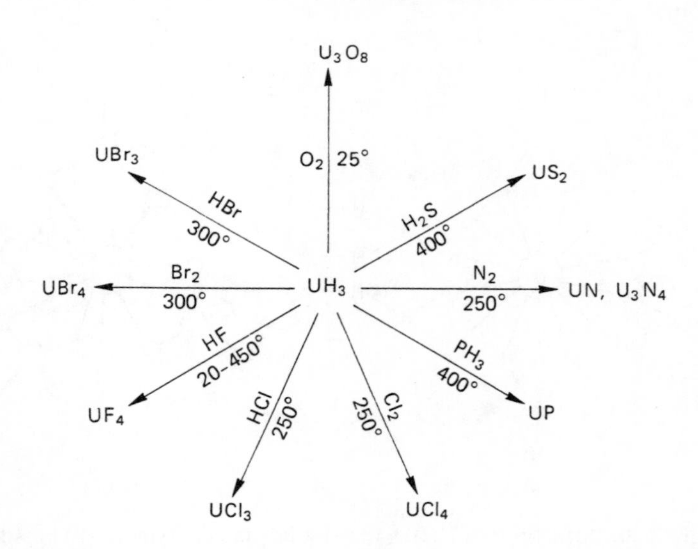

3.16 Chemistry of the elements – neptunium

This was the first transuranium element to be identified, in 1940. Of the 15 known isotopes, only ^{237}Np ($t_{1/2} = 2.14 \times 10^6$ y) combines a long half-life with availabilty in sufficient quantity for practical work. Compared with its flanking elements (U and Pu) it has received relatively little study. Obtained mainly as a by-product of uranium and plutonium processing, it exhibits oxidation numbers from (III) to (VII) in its compounds.

The metal

Neptunium can be obtained by calcium reduction of NpF_4 (using iodine booster):

$$NpF_4 + 2Ca \xrightarrow{1200°} Np + 2CaF_2$$

It is a silvery metal, with a density of 19.5 g cm^{-3}, melting point of 637° and estimated boiling point of 4174°. Generally similar to uranium, it undergoes surface oxidation by air, being wholly converted to NpO_2 at high temperatures.

The oxidation states of neptunium

Neptunium marks a transition between uranium and plutonium. Thus the +6 oxidation, attained in NpF_6 and the NpO_2^{2+} ion, is stabler than Pu(VI) but not as stable as UO_2^{2+}. In contrast to UO_2^+, NpO_2^+ is stable to disproportionation; it is, however, reduced rapidly to Np^{4+} by Fe^{2+}.

Neptunium(IV) is made by reduction of higher oxidation states and by aerial oxidation of Np^{3+}. It is oxidised back to NpO_2^{2+} by Ce^{4+} for example. Neptunium(III) is, in contrast to U(III), stable in the absence of air; it may conveniently be obtained by electrolytic reduction. The oxidation of Np(VI) to Np(VII) may be accomplished, for example, by ozone (in 1M alkali) at room temperature, or by other oxidising agents (XeO_3, IO_4^-) at higher temperatures.

The relevant redox potentials are:

(1M acid)

$$Np \xrightarrow{1.79\text{ V}} Np^{3+} \xrightarrow{-0.15\text{ V}} Np^{4+} \xrightarrow{-0.67\text{ V}}$$
$$NpO_2^+ \xrightarrow{-1.24\text{ V}} NpO_2^{2+}$$

(1M alkali)

$$NpO_2(OH)_2 \xrightarrow{-0.6\text{ V}} NpO_5^{3-}$$

Neptunium oxides

Neptunium forms oxides NpO_2 and Np_2O_5 as well as hydrated forms of $NpO_3.xH_2O$ ($x = 1, 2$) which may be hydroxides of the type $NpO_2(OH)_2$. The apple green NpO_2 is the usual product of ignition of neptunium compounds in air at 700–800°, and has the UO_2 (fluorite) structure. The preparation of the higher oxides is difficult, unlike those of uranium.

NpO_2 reacts with many metal oxides on heating to form mixed oxides, usually involving Np(VI). Several, such as $BaNpO_4$ and K_2NpO_4, have been shown to contain NpO_2^{2+} groups. Latterly, some neptunates (VII) have been made, usually by heating the metal oxides together in a stream of air or oxygen:

$$Li_2O + NpO_2 \xrightarrow[400^\circ]{O_2} Li_5NpO_6$$

Neptunium(VII) is stable in very basic solutions, as NpO_5^{3-} is stable in salts with suitable counter ions (such as Ba^{2+}, $Co(NH_3)_6^{3+}$).

Neptunium halides

All four halides NpX_3 exist, as well as NpX_4 (X = F, Cl, Br), NpF_5 and NpF_6. Syntheses include:

$$NpO_2 \xrightarrow[500^\circ]{H_2/HF} NpF_3 \text{ (purple)}$$

$$NpCl_4 \xrightarrow{H_2,\ 450^\circ} NpCl_3 \text{ (green)}$$

$$NpO_2 \xrightarrow[500^\circ]{HX\ (X = Br,\ I)} NpX_3 \text{ (X = Br, green; X = I, brown)}$$

$$NpO_2 \xrightarrow{HF,\ 500^\circ} NpF_4 \text{ (green)}$$

$$NpO_2 \xrightarrow[500^\circ]{CCl_4} NpCl_4 \text{ (red-orange)}$$

$$NpO_2 \xrightarrow{AlBr_3,\ 350^\circ} NpBr_4 \text{ (dark red)}$$

$$NpF_4 \xrightarrow{KrF_2/HF} NpF_5 \text{ (blue-white)}$$

$$NpF_3,\ NpF_4,\ NpO_2 \xrightarrow[500^\circ]{F_2} NpF_6 \text{ (orange)}$$

All have the same structures as the analogous uranium compounds, except that there is a second form of $NpBr_3$ with the $PuBr_3$ structure. Like its uranium and plutonium analogues, NpF_6 is volatile (as a red-brown gas, bp 55.2°) and toxic.

It might be expected that NpF_7 would be isolable, but attempts using plausible fluorinating agents (such as KrF_2) have been unsuccessful. In view of the stability of Np(IV), the non-existence of NpI_4 is perhaps surprising, but parallels can be found in 3*d* chemistry in the instability of FeI_3 and CuI_2.

A few oxohalides – $NpOF_3$, NpO_2F_2, $NpOF_4$, NpOCl, NpOI – and complex halides exist. Na_3NpF_8 is isostructural with Na_3UF_8, Rb_2NpF_7 has 7-coordinate Np (capped trigonal prism) while $CsNpF_6$ and Cs_2NpCl_6 have octahedrally coordinated neptunium, as for the uranium analogues.

Other binary compounds

These are synthesised by direct combination of the elements as well as by other methods; examples include:

$$Np \xrightarrow{H_2/\text{heat}} NpH_2, NpH_3$$

$$Np \xrightarrow[740^\circ]{P} Np_3P_4$$

$$NpO_2 \xrightarrow[2600^\circ]{C} NpC_2$$

$$NpH_3 \xrightarrow[1430^\circ]{C} NpC$$

$$NpH_3 \xrightarrow[800^\circ]{NH_3} NpN$$

$$NpF_3 \xrightarrow[1500^\circ]{Si} NpSi_2$$

They are generally isostructural with other actinide analogues – $NpSi_2$ has the $ThSi_2$ structure; NpC and NpN (mp 2830°) are isostructural with the uranium and plutonium analogues.

Neptunium complexes

These are mainly known for the +4 and +6 oxidation states. X-ray diffraction studies have been made on several, such as the 8-coordinate $NaNpO_2(CH_3CO_2)_3$, $Na_4NpO_2(O_2)_3.9H_2O$, $NpO_2(NO_3)_2.6H_2O$, $K_4NpO_2(CO_3)_3$ and $KNpO_2CO_3$. These are isostructural with the corresponding compounds of U, Pu and Am. Besides confirming the existence of the NpO_2^{2+} (neptunyl) moiety, these studies demonstrate an incremental contraction of 0.01 Å in M–O distance proceeding down the series of U, Np, Pu and Am isostructural analogues. $NpO_2(NO_3)_2$.bipy is also 8-coordinate while $NpO_2(acac)_2$.py is 7-coordinate. Study of

$BaNpO_2(CH_3CO_2)_3$ shows a lengthening of 0.14 Å in the Np–O bond in going from the Np(VI) to the Np(V) compound, a consequence of an electron being added to an antibonding orbital.

Eight coordination not involving neptunyl groups is found in $Np(S_2CNEt_2)_4$, $(Et_4N)_4Np(NCS)_8$, and $Np(acac)_4$ (respectively dodecahedral, cubic and antiprismatic). In contrast, $Np(BH_4)_4$ resembles $Pu(BH_4)_4$ (rather than $U(BH_4)_4$) in having a monomeric structure with tridentate BH_4 groups affording 12-coordination.

Various halide complexes can be isolated, like $NpX_4(R_3PO)_2$ (X = Cl, Br; R = Ph, NMe_2), $NpCl_4(Me_3PO)_2$ and $NpCl_4(Me_2SO)_3$, as well as the related $Np(NCS)_4(R_3PO)_4$ (R = Ph, Me, Me_2N) and $Np(NO_3)_4(Me_2SO)_3$. These display a variety of hue – the chlorides and bromides are typically pink or green, and the thiocyanates yellow; where structures are known, they are isostructural with the corresponding thorium and uranium compounds.

Few neptunium(III) compounds have been characterised. One such is $Et_4N^+[Np(S_2CNEt_2)_4]^-$ whose distorted dodecahedral 8-coordination is paralleled in similar compounds of plutonium and the lanthanides.

3.17 Chemistry of the elements – plutonium

Despite its toxicity and in contrast to the metals immediately preceding and succeeding it, this is an intensively studied element, because of its potential use as a fuel and in weaponry.

Although 15 isotopes are known, with masses from 232 to 246, the most important is the fissionable ^{239}Pu. Its half-life of 24 100 years makes it suitable for chemical study; apart from its toxicity, the possibility of its critical mass being exceeded is a potential problem for any large-scale study.

Plutonium exhibits oxidation states from 3 to 7 in its compounds; oxidation states of 3 and 4 are more important (compared with U and Np) but compounds of the plutonyl ion, analogous to UO_2^{2+}, are well-defined. Pu(VII) only exists under very alkaline conditions.

The metal

This may be made by various routes; commercially usually involving metallothermic (Ca) reduction of PuF_4, PuF_3, $PuCl_3$ or $CaPuF_6$. Plutonium is unusual in existing in six allotropic forms at normal pressure between room temperature and its melting point (640°). It is a rather dense (19.8 g cm^{-3}) silvery and reactive metal; more reactive than uranium or neptunium, it is attacked by dry air, forming a grey-green coating of the oxide. It reacts very slowly with cold water, rather faster with dilute H_2SO_4, and dissolves quickly in dilute HCl or HBr. It is passivated by nitric acid but does dissolve in boiling acid in the presence of HF, doubtless owing to the formation of complexes like PuF_6^{2-}.

Plutonium oxides

The Pu–O phase diagram is complex. Stoichiometric phases Pu_2O_3 (*A*-type La_2O_3 structure) and PuO_2 (fluorite structure) exist. The former is made by reduction of PuO_2, the latter by heating many Pu compounds in air:

$$3PuO_2 + Pu \xrightarrow{1500^\circ} 2Pu_2O_3$$

$$Pu(C_2O_4)_2 \xrightarrow{870^\circ} PuO_2 + 2CO + 2CO_2$$

$$Pu(NO_3)_4 \xrightarrow{870^\circ} PuO_2 + 4NO_2 + O_2$$

As with uranium, there are many mixed oxides. These are formed when mixtures of PuO_2 and metal oxides, peroxides and carbonates are heated, usually in a stream of oxygen. Oxidation numbers of III to VII are known, with the hexavalent state most common:

$$4Li_2O + PuO_2 \xrightarrow[600^\circ]{\text{sealed tube}} Li_8PuO_6$$

$$2Li_2O + PuO_2 + \tfrac{1}{2}O_2 \xrightarrow{400\text{–}500^\circ} Li_4PuO_5 \xrightarrow[1000^\circ]{700\text{–}} Li_3PuO_4$$

$$3CaO + PuO_2 + \tfrac{1}{2}O_2 \xrightarrow{950\text{–}1050^\circ} Ca_3PuO_6$$

$$SrO + PuO_2 + \tfrac{1}{2}O_2 \xrightarrow{900\text{–}1000^\circ} SrPuO_4$$

$$5Li_2O + 2PuO_2 + \tfrac{3}{2}O_2 \xrightarrow{430^\circ} 2Li_5PuO_6$$

In general, the Pu(VI) compounds are isomorphous with U(VI) analogues; thus $SrPuO_4$ has, like $SrUO_4$, PuO_2^{2+} ions surrounded by four oxygen neighbours. The oxoplutonates (VII), such as M_3PuO_5 (M = Rb, Cs), Li_5PuO_6 and $Ba(PuO_5)_2.xH_2O$, have not yet been structurally characterised.

Plutonium(VII) is stable in solution only under very alkaline conditions; below $[OH^-]$ of 7M, it oxidises water to oxygen.

Other binary compounds

These are formed with a wide range of non-metals; some representative preparations are shown in Figure 3.31, many having been studied as potential fuels. The hydrides are sources of metallic plutonium as at 400°

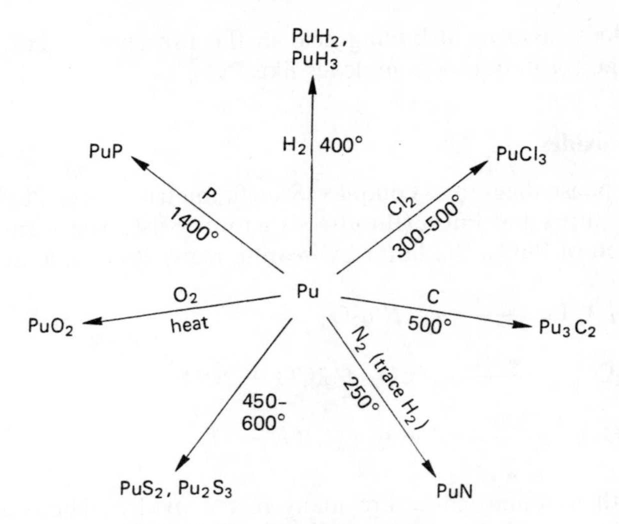

Figure 3.31 Reactions of plutonium

they afford synthetically useful metal powder. The reason for the trace of hydrogen in the preparation of PuN is that it acts as a catalyst via the formation of plutonium hydride.

Plutonium halides

All the plutonium halides PuX_3 (X = F, Cl, Br, I) exist and are generally similar to the corresponding uranium compounds. The only others isolated are PuF_4 and PuF_6, demonstrating the trend in increasing instability of the higher oxidation states in the actinides after uranium. Syntheses include:

$$2PuO_2 + H_2 + 6HF \xrightarrow{600^\circ} 4H_2O + \underset{\left(\begin{array}{c}\text{violet-blue}\\ \text{solid}\end{array}\right)}{2PuF_3}$$

$$Pu_2(C_2O_4)_3.10H_2O + 6HCl \longrightarrow 3CO + 3CO_2 + 13H_2O + \underset{\left(\begin{array}{c}\text{blue-green}\\ \text{solid}\end{array}\right)}{2PuCl_3}$$

$$PuH_3 + 3HBr \xrightarrow{600^\circ} 3H_2 + \underset{\left(\begin{array}{c}\text{light green}\\ \text{crystals}\end{array}\right)}{PuBr_3}$$

$$2Pu + 3HgI_2 \xrightarrow{500^\circ} 3Hg + \underset{\left(\substack{\text{bright green}\\ \text{solid}}\right)}{2PuI_3}$$

$$Pu(aq) + HF(aq) \longrightarrow PuF_4.2\tfrac{1}{2}H_2O(s) \xrightarrow[\text{heat}]{HF(g)} \underset{\left(\substack{\text{pale-brown}\\ \text{solid}}\right)}{PuF_4}$$

$$PuF_4 + F_2 \xrightarrow{400^\circ} \underset{\left(\substack{\text{red brown}\\ \text{solid}}\right)}{PuF_6}$$

Despite the instability of the +6 state, PuF_6 has been isolated, as a very volatile (and toxic) solid (mp 51.6°; bp 62.15°). It loses fluorine readily and tends to be decomposed by its own α-radiation; it is best stored in the gas phase when it resembles NO_2. It has potential for plutonium recovery.

$PuCl_4$ does not exist in the solid state but has been detected by its absorption spectrum in the vapour phase above 900°C, as the equilibrium

$$PuCl_3 + \tfrac{1}{2}Cl_2 \rightleftharpoons PuCl_4$$

moves to the right on heating. Hexachloride salts, $PuCl_6^{2-}$, are well-defined, as are Lewis base adducts $PuCl_4.L_x$ ($x = 1, 2$), discussed later.

A few oxyhalides exist: PuOX (X = F, Cl, Br, I), $PuO_2Cl_2.6H_2O$, $PuOF_4$ and PuO_2F_2, the last two produced by controlled low-temperature hydrolysis of PuF_6. The absence of compounds in the +4 and +5 and oxidation states should again be noted.

Halide complexes

These are best-known for Pu(IV) and generally resemble the corresponding uranium complexes. No iodide complexes of Pu(IV) appear to be known. Preparations include:

$$LiF + PuF_6 \xrightarrow{300^\circ} LiPuF_5 \xrightarrow{400^\circ} Li_4PuF_8$$

$$PuF_6 \xrightarrow[300^\circ]{CsF} F_2 + Cs_2PuF_6 \xrightarrow[>300^\circ]{F_2} CsPuF_6$$

$$PuF_4 \xrightarrow[70\text{–}100^\circ]{NH_4F} NH_4PuF_5 + (NH_4)_2PuF_6$$

$$Pu^{4+}(aq) \xrightarrow[\text{conc. HCl}]{CsCl} Cs_2PuCl_6$$

$$Pu^{4+}(aq) \xrightarrow[Et_4NBr]{\text{conc. HBr/EtOH}} (Et_4N)_2PuBr_6$$

Some of these complexes feature 6-coordinate Pu, established for $CsPuF_6$ and highly probably for Cs_2PuCl_6 and $(Et_4N)_2PuBr_6$; like the 9-coordinate $(NH_4)_4PuF_8$, many fluorides have higher coordination numbers. $CaPuF_6$ is used in the preparation of plutonium by reduction with Ca.

The chloride and bromide complexes are useful potential starting materials for the synthesis of PuX_4 (X = Cl, Br) complexes which cannot, of course, be made directly, though the bromides are usually made by oxidation of $PuBr_3$ complexes:

$$Cs_2PuCl_6 \xrightarrow[Me_2SO]{\text{exc. hot}} PuCl_4(Me_2SO)_7 \xrightarrow[MeNO_2]{MeCN} PuCl_4(Me_2SO)_3$$

$$Cs_2PuCl_6 \xrightarrow[MeCN]{Ph_3PO} PuCl_4(Ph_3PO)_2$$

$$PuBr_3 \xrightarrow[Ph_3PO]{Br_2/MeCN} PuBr_4(Ph_3PO)_2$$

Many such complexes have been isolated, such as $PuCl_4(Ph_2SO)_x$ (x = 3, 4), *trans*-$PuX_4[(Me_2N)_3PO]_2$ (X = Cl, Br) and $PuCl_4(Me_3PO)_6$ (probably $PuCl(Me_3PO)_6^+(Cl^-)_3$).

Similar thiocyanate complexes are obtained by metathesis:

$$PuCl_4L_2 \xrightarrow[Me_2CO/MeOH]{\text{KNCS/exc. L}} Pu(NCS)_4L_4 + 4KCl$$

(L = Ph_3PO, $(Me_2N)_3PO$)

The complex can be crystallised, after filtering off the KCl and evaporating the solvent.

Other complexes

The oxalates – $Pu_2(C_2O_4)_3.xH_2O$ (x = 10, 11), $Pu(C_2O_4)_2.\ 6H_2O$ and $PuO_2(C_2O_4)_2.3H_2O$ – all precipitate on addition of oxalic acid solution to solutions of the corresponding plutonium ions. Their insolubility has been utilised in plutonium reprocessing: on heating in air, they all yield PuO_2.

Carbonate complexes are important in the transport of Pu in natural waters and also because they can be employed in the preparation of plutonium ceramics:

$$PuO_2^{2+}(aq) \xrightarrow{Na_2CO_3(aq)} PuO_2CO_3(s) \xrightarrow[Na_2CO_3]{\text{exc.}} [PuO_2(CO_3)_3]^{4-}(aq)$$

The ammonium salt decomposes in two stages on heating, allowing isolation of PuO_2CO_3:

$$(NH_4)_4PuO_2(CO_3)_3 \xrightarrow[150°]{110-} 4NH_3 + 2H_2O + 2CO_2 + PuO_2CO_3$$

$$PuO_2CO_3 \xrightarrow{250\text{–}300°} PuO_2 + CO_2 + \tfrac{1}{2}O_2$$

Nitrate complexes are important in the separation chemistry of plutonium (section 3.8). Dark green crystals of $Pu(NO_3)_4.5H_2O$ are obtained from solutions of Pu(IV) in concentrated nitric acid; they contain 11-coordinate $Pu(NO_3)_4.(H_2O)_3$ molecules. In very concentrated (10–14M) HNO_3, $Pu(NO_3)_6^{2-}$ ions exist, which may be isolated as yellow salts $M_2Pu(NO_3)_6$ (M = Cs, Et_4N). Plutonium(VI) in concentrated HNO_3 affords red-brown to purple crystals of $PuO_2(NO_3)_2.6H_2O$, isomorphous with the U(VI) analogue, and thus having 8-coordinate $PuO_2(NO_3)_2(H_2O)_2$ molecules.

Similar 8-coordinate pink plutonyl complexes $M^+PuO_2(CH_3COO)_3^-$ (M = Cs, Na, for example) are precipitated by adding acetate ions to a Pu(VI) solution in the presence of the alkali metal. 8-coordination is found in the neutral Pu(IV) complexes PuL_4 (L = acac, hfa, 8-hydroxyquinoline, tropolonate), believed to have the same structures as the uranium analogues; such neutral complexes are potentially useful in the solvent extraction of plutonium.

The borohydride $Pu(BH_4)_4$ is a blue-black volatile liquid (mp 15°); isostructural with the Np analogue, it has four tridentate BH_4 groups and thus 12-coordinate plutonium.

Plutonium aquo ions

The blue Pu^{3+}(aq) ion has an acidity similar to tripositive lanthanide ions, while the brown Pu^{4+}(aq) only exists in strong acid solution; at higher pH values it tends to polymerise, eventually precipitating plutonium(IV) hydroxide, probably hydrated PuO_2. The purple-pink PuO_2^+ tends to disproportionate; PuO_2^{2+} is orange-yellow; the blue Pu(VII) ion, possible PuO_5^{3-}(aq) only exists in very strong alkali. The existence of dioxoplutonium species is confirmed by their isolation in crystalline compounds as well as by the observation of PuO_2 stretching vibrations as fine structure in the electronic absorption spectra and in the IR spectra of plutonyl compounds.

The stability of plutonium oxidation states

In aqueous solution, two effects are at work. One is the general tendency of oxidation states IV–VI to be reduced owing to the effects of α-decay (1 mg ^{239}Pu emits over a million α-particles a second); the second is the possibility of disproportionation and reproportionation reactions.

The redox potentials for plutonium are as follows:

$$Pu \xrightarrow{2.03\ V} Pu^{3+} \xrightarrow{-0.986\ V} Pu^{4+} \xrightarrow{-1.17\ V} PuO_2^+ \xrightarrow{-0.916\ V} PuO_2^{2+}$$

The Pu(III)–Pu(IV) and Pu(V)–Pu(VI) couples are reversible, but not Pu(IV)–Pu(V), as this involves making and breaking Pu–O bonds; an analogous situation applies in the case of uranium. The potentials separating the four common oxidation states are roughly equal. Thus the disproportionation of Pu(IV) in aqueous solution:

$$2Pu^{4+} + 2H_2O \longrightarrow Pu^{3+} + PuO_2^+ + 4H^+$$

made up of the processes:

$$Pu^{4+} + e^- \longrightarrow Pu^{3+}$$

$$Pu^{4+} + 2H_2O \longrightarrow PuO_2^+ + 4H^+ + e^-$$

takes place with a relatively small free energy change (< 20 kJ mol^{-1}).

This may be followed by:

$$2PuO_2^+ + 2H^+ \longrightarrow PuO_2^{2+} + Pu^{4+} + H_2O$$

Superimposed on this process is autoradiolysis – the reduction of the higher oxidation states to Pu(IV) by the effects of plutonium's own α-radiation.

In acid solution, both decomposition of the solvent (to H_2O_2) and acid occur; in HNO_3, both HNO_2 and nitrogen oxides are formed. Thus successively:

$$PuO_2^{2+} + e^- \longrightarrow PuO_2^+$$

$$PuO_2^+ + e^- + 4H^+ \longrightarrow Pu^{4+} + 2H_2O$$

$$Pu^{4+} + e^- \longrightarrow Pu^{3+}$$

and also disproportionation and reproportionation occur:

$$2PuO_2^+ + 4H^+ \longrightarrow PuO_2^{2+} + Pu^{4+} + 2H_2O$$

$$PuO_2^+ + Pu^{3+} + 4H^+ \longrightarrow 2Pu^{4+} + 2H_2O$$

The combined effect of the reduction due to autoradiolysis and of disproportionation is shown in Figure 3.32.

3.18 Chemistry of the elements – americium

This element was first synthesised in 1944–45 by Seaborg and coworkers; neutron irradiation of ^{239}Pu afforded ^{241}Pu, which decayed to ^{241}Am (433 y). Along with ^{243}Am (7380 y), these are the most important of the 12 known isotopes; both are available on the kg scale.

The metal is made by metallothermic reduction:

$$AmF_3 \xrightarrow[1000\text{–}1200°]{\text{Ba or Li}} Am$$

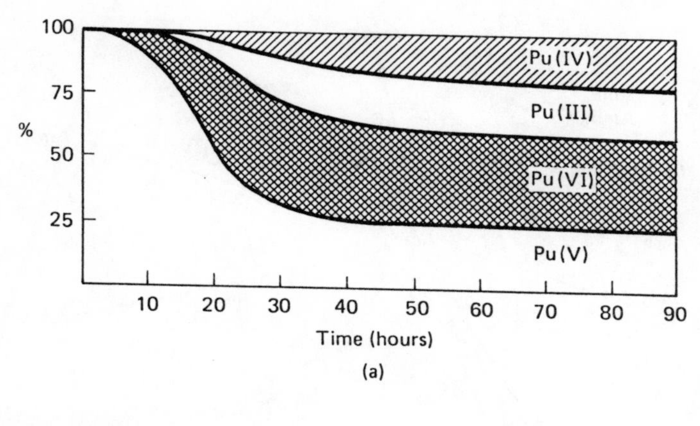

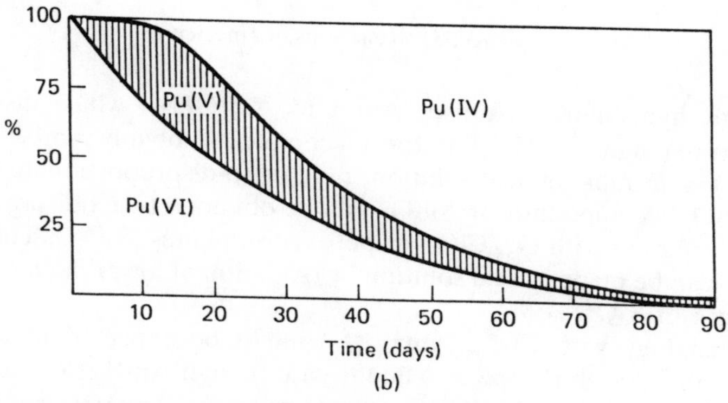

Figure 3.32 (a) Disproportionation of Pu(V) in 0.1 N HNO_3 + 0.2 N $NaNO_3$; (b) self-reduction of Pu(VI) due to its α-emission, in 5 N $NaNO_3$ [after P. I. Artiukhin, V. I. Medcedovskii and A. D. Gel'man, *Radiokhimiya*, **1** (1959) 131; *Zh. Neorg. Khim.*, **4** (1959) 1324]

$$3AmO_2 + 4La \longrightarrow 2La_2O_3 + 3Am$$

Owing to the lesser volatility of lanthanium (bp 3457°), americium may be obtained in a very pure state (99.9 per cent) by the latter method. Americium is a silvery, ductile, very malleable metal (mp 994°, bp 2607°). It tarnishes slowly in moist air and dissolves readily in dilute HCl, as would be expected from its electropositive character ($E(Am/Am^{3+})$ = 2.36 V); the metal reacts on heating with oxygen, the halogens and many non-metals (Figure 3.33).

Oxidation states of americium

Following the trend noticeable through plutonium, the pink Am^{3+} is normally the stable oxidation state in aqueous solution, Am(IV) being unstable in the absence of complexing anions. In alkaline solution,

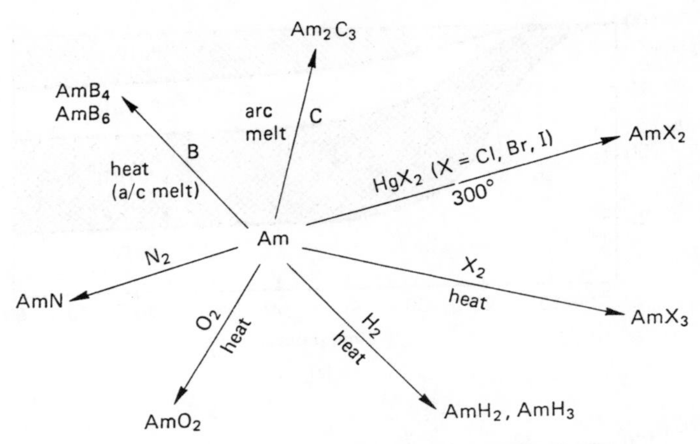

Figure 3.33 Reactions of americium

however, hypochlorite oxidises Am^{3+} to $Am(OH)_4$, which dissolves in concentrated aqueous NH_4F to form a complex, probably AmF_8^{4-}. Americium(V) is unstable in acid solution, tending to disproportionate to Am^{3+} and AmO_2^{2+}; compounds of AmO_2^+ can be obtained by oxidising Am^{3+} in alkaline solution with O_3, ClO^- or peroxydisulphates, $S_2O_8^{2-}$. The related AmO_2^{2+} can be made in acid solution by oxidation of lower oxidation states with Ag^{2+} or $S_2O_8^{2+}$.

By analogy with Eu^{2+}, Am^{2+} (f^7) might be expected to obtain in solution; this does not appear to be the case though AmX_2 (X = Cl, Br, I) exist in the solid state. At the other extreme, some Am(VII) can be made from alkaline solutions of Am(VI).

As with plutonium, there are two factors affecting the stability of oxidation states – Am(IV) and (V) tend to disproportionate:

$$2Am(IV) \longrightarrow Am(III) + Am(V)$$

and

$$Am^{4+} + AmO_2^+ \longrightarrow AmO_2^{2+} + Am^{3+}$$

also

$$3AmO_2^+ + 4H^+ \longrightarrow 2AmO_2^{2+} + Am^{3+} + 2H_2O$$

Superimposed on this, α-radiolysis tends to ensure that higher oxidation states are reduced to Am^{3+}.

Oxides

There are two well-characterised oxides, Am_2O_3 and AmO_2. The black AmO_2 (fluorite structure like other AnO_2) is made by pyrolysis of various Am salts (carbonate, nitrate, oxalate) at 600–800°. Hydrogen reduction of

AmO_2 at 600° yields the red-brown, air-sensitive Am_2O_3, which exhibits all three rare-earth sesquioxide structures, the C-type (bcc) being stable at room temperature.

Halides and halide complexes

Americium is the first actinide to form dihalides:

$$Am + HgX_2 \xrightarrow{\text{heat}} AmX_2 + Hg \ (X = Cl, Br, I)$$

They are black solids with the $PbCl_2$, $EuBr_2$ and EuI_2 structures respectively. As expected, americium forms the full range of trihalides and a tetrafluoride:

$$AmO_2 \xrightarrow{HF/HNO_3} AmF_4.xH_2O \xrightarrow[125°]{NH_4HF_2} \underset{\text{(pink)}}{AmF_3}$$

$$AmO_2 \xrightarrow[800°]{CCl_4} \underset{\text{(pink)}}{AmCl_3}$$

$$AmO_2 \xrightarrow[500°]{AlBr_3} \underset{\left(\text{white to pale yellow}\right)}{AmBr_3}$$

$$AmO_2 \xrightarrow[500°]{AlI_3} \underset{\text{(pale yellow)}}{AmI_3}$$

$$AmCl_3 \xrightarrow[400°]{NH_4I} AmI_3$$

$$AmF_3 \xrightarrow{F_2} \underset{\text{(tan)}}{AmF_4}$$

This resemblance to plutonium is deepened by the claims for a hexafluoride, AmF_6, as a volatile dark brown solid (IR ν(Am–F) 604 cm^{-1}):

$$AmF_4 \xrightarrow{KrF_2/HF,\ 40\text{–}60°} AmF_6$$

Several halide complexes have been characterised: thus $(NH_4)_4AmF_8$ is 8-coordinate, like the uranium analogue. In the +3 oxidation state, the halide complexes $Cs_2NaAmCl_6$ and $(Ph_3PH)_3AmX_6$ (X = Cl, Br) are doubtless octahedral (the oxyhalides AmO_2F_2 and complexes $M_2AmO_2Cl_4$ (M = Rb, Cs) and $Cs_2AmO_2Cl_4$ have been reported).

Other complexes

Compared with the earlier actinides, they have been little investigated. The AmO_2^+ and AmO_2^{2+} ions are well-defined; in HCl solution, the symmetric stretching vibration is found at $796\ cm^{-1}$ (Am(VI)) and $730\ cm^{-1}$ (Am(V)). The X-ray study of the lemon coloured $NaAmO_2(OCOCH_3)_3$ shows it to have the 8 (2 + 6) coordination found for the U, Np and Pu analogues. Eight-coordination is also established for AmO_2F_2, $MAmO_2F_2$ (M = Rb, K) and $MAmO_2CO_3$ (M = K, Rb, Cs). $Cs_2AmO_2Cl_4$ has 6-coordination as in $UO_2Cl_4^{2-}$, and $NH_4AmO_2PO_4$ is also 6-coordinate. There is a considerable chemistry of Am(III), in contrast with, say, uranium; thus $Am_2(SO_4)_3.8H_2O$ is isomorphous with 8-coordinate lanthanide analogues (M = Pr–Sm), while the pink $Am_2(C_2O_4)_3.xH_2O$ (x = 7, 11) have also been isolated. There are several diketonates: $Am(Me_3CCOCHCOCMe_3)_3$ is a 7-coordinate dimer like the praseodymium analogue; $Am(CF_3COCHCOCF_3)_3[(BuO)_3PO]_2$ is volatile at 175°, such volatility being a potential route to the separation of actinides.

3.19 Chemistry of the actinides beyond americium (Cm, Bk, Cf, Es, Fm, Md, No, Lr)

These elements differ from the preceding actinides in a number of respects. They mark the point where lanthanide-like behaviour becomes increasingly the norm; thus their chemistry is mainly that of the M(III) oxidation state, there being no evidence for the MO_2^+ or MO_2^{2+} ions prominent for U–Am. The parallels are not exact; while in the lanthanide series, ions with filled or half-filled f shells are particularly stable, compounds of Cf^{4+} (f^8), Cf^{2+} (f^{10}), Es^{2+} (f^{11}) and Fm^{2+} (f^{12}) are as well-defined as those of Bk^{4+} (f^7) and No^{2+} (f^{14}). Another significant difference between the two series is that the later lanthanides form much more stable chloride complexes, a fact important in the separation of lanthanide fission products from actinides. The similarity between the later actinides, owing to their adoption of the 3+ oxidation state with consequential similarity in ionic radii and chemistry, means that ion-exchange or solvent-extraction methods are used to effect their separation, just as they are within the lanthanide series.

A considerable number of solid compounds have been isolated and characterised for Cm, Bk, Cf and to a lesser extent Es – mainly by using methods outlined in section 3.9; in general, they have been binary compounds like the trihalides, along with certain other halides, oxyhalides, and compounds with chalcogenides and pnictides. From fermium onwards, the knowledge of the chemistry of the elements has been derived from tracer studies.

Table 3.8 reveals some patterns in the compounds of these elements that have been made; perusal suggests compounds that might be expected to exist (such as BkF_6^{2-}, $CmHal_6^{3-}$). Study of the redox potentials (Table 3.2) shows that, after Am, Bk is the most likely to form stable compounds

Table 3.8 Some isolated compounds of the trans-plutonium elements

	Am	Cm	Bk	Cf	Es
An^{2+}	f^7	f^8	f^9	f^{10}	f^{11}
An^{3+}	f^6	f^7	f^8	f^9	f^{10}
An^{4+}	f^5	f^6	f^7	f^8	f^9
Tetrahalides	AmF_4	CmF_4	BkF_4	CfF_4	EsF_4?
Trihalides	All	All	All	All	All
Dihalides	AmX_2	—	—	$CfBr_2$, CfI_2	EsX_2
Oxyhalides	AmOCl, AmOBr	CmOCl	BkOX	All	EsOX
Oxides	AmO_2, Am_2O_3	CmO_2, Cm_2O_3	BkO_2, Bk_2O_3	CfO_2, Cf_2O_3	Es_2O_3
*$MHal_6^{2-}$	AmF_6^{2-}	CmF_6^{2-}	$BkCl_6^{2-}$		
*$MHal_6^{3-}$	AmF_6^{3-}		$BkCl_6^{3-}$		
*MQ	AmP, AmSb AmB	CmN, CmP CmAs, CmSb	BkN, BkP BkAs, BkSb	CfN, CfAs CfSb	

*Hal = Cl, Br, I. Q = pnictide.

in the (IV) oxidation state, which increasingly grows less stable. Conversely, the (II) oxidation state becomes increasingly stable as the atomic number increases.

The elements and their reactions

Curium, berkelium and californium have all been isolated, typically by reduction of the fluoride with lithium or of the oxide with lanthanum or thorium. They are quite electropositive metals, oxidised by air to the oxide, reacting with hydrogen on warming to form hydrides, and yielding compounds on warming with group V and VI non-metals. Synthesis of compounds of the trans-americium elements over the last 25 years has demanded considerable experimental expertise; here are some case studies which put into practice points made in section 3.9.

Curium(III) chloride

The original study of this compound reported in 1965–67 was complicated by self-irradiation of the ^{244}Cm ($t_{1/2}$ = 18.1 y). By 1970 the longer lived ^{248}Cm ($t_{1/2} = 3.4 \times 10^5$ y) was available on the microgram scale; ion-exchange beads, each loaded with a μg of the isotope, were calcined at 1200° to afford Cm_2O_3, which was loaded into capillaries and treated with HCl at 500° forming $CmCl_3$. A single crystal of this very hygroscopic pale yellow solid was made by slow cooling of molten $CmCl_3$ from above 600° under a HCl flow; X-ray diffraction on this crystal confirmed the adoption of the UCl_3 structure and yielded the Cm–Cl bond lengths.

Californium(III) chloride and californium(II and III) iodides

The original report (1973) of the structure of $CfCl_3$ was carried out using a sample prepared by the same method as $CmCl_3$. Subsequently, however, the availability of ^{249}Cf (351 y) on the μg scale permitted an improved method.

Californium oxalate was initially precipitated in the mouth of a quartz capillary and filtered by air pressure in the tube using a plug of ashless filter paper. Heating at 800–900° (using a tube furnace) turned the oxalate into Cf_2O_3 (and destroyed the filter paper). Treatment of the oxide with CCl_4 at 520–550° afforded lime-green $CfCl_3$, whose identity was confirmed by its X-ray diffraction pattern. After purification by sublimation at 800°, the $CfCl_3$ was used as a starting material for further synthesis:

$$CfCl_3 \xrightarrow[540^\circ]{HI(g)} \underset{\left(\begin{array}{c}\text{yellow-}\\\text{orange}\end{array}\right)}{CfI_3} \xrightarrow[520\text{–}570^\circ]{H_2} \underset{\left(\begin{array}{c}\text{lavender}\\\text{violet}\end{array}\right)}{CfI_2}$$

These iodides were characterised by X-ray diffraction and by UV–visible spectroscopy.

Einsteinium(III) chloride

The initial synthesis was achieved (1968) despite several handicaps. The short-lived ^{253}Es ($t_{1/2}$ = 20 d) releases some 1400 kJ mol^{-1} min^{-1}, sufficient to dislocate a crystal lattice within 3 minutes of synthesis and also capable of destroying ion-exchange resins. Some 200 nanograms of einsteinium were absorbed on to a small chip of coconut charcoal, transferred to a capillary, and converted successively to Es_2O_3 and $EsCl_3$ in the manner described above. The X-ray diffraction study was carried out just below the melting point of $EsCl_3$ so that radiation-induced defects were removed by annealing.

By 1978, a different route was chosen for the study of EsF_3, carried out on the microgram scale:

$$Es^{3+}(aq) \xrightarrow{C_2O_4^{2-}(aq)} Es_2(C_2O_4)_3(s) \xrightarrow{1000^\circ} Es_2O_3(s) \xrightarrow[(2)\ F_2/300^\circ]{(1)\ HF} EsF_3(s)$$

By studying the absorption spectrum of the EsF_3 over a period of up to 2 years, it was possible to observe the spectrum of both the $^{249}BkF_3$ daughter and $^{249}CfF_3$ granddaughter, and assign them to compounds with the LaF_3 structure.

Studies on the trans-einsteinium elements

For the elements beyond einsteinium, particular use has been made of ion-exchange and solvent-extraction techniques, allied with coprecipitation studies to clarify the oxidation states of the actinide ions in solution.

Thus fermium coprecipitates with lanthanide fluorides and hydroxides, confirming it to form Fm^{3+} ions. It elutes from cation-exchange resins just before Es^{3+} and its complexes (with chloride and thiocyanate) are eluted from anion-exchange resins just after einsteinium; hence Fm^{3+} forms slightly stronger complexes than Es^{3+}, as expected for an ion with smaller ionic radius.

Tracer studies of nobelium in solution have, however, indicated un-actinide-like behaviour. Nobelium ions (No^{2+}) precipitate with BaF_2 rather than with LaF_3, but after adding Ce^{4+} oxidant to the nobelium, half the nobelium precipitates with LaF_3, showing it to have been oxidised to No^{3+}.

The pattern of alkaline-earth-like behaviour is followed in complexing studies with Cl^-; elution behaviour on a cation-exchange resin and solvent-exchange studies resemble Ca^{2+}. Elution from a cation-exchange resin using first 0.3M then 1.9M ammonium α-hydroxyisobutyrate as eluent found that Es^{3+} tracer, but not nobelium, was eluted by the more dilute solution; with the more concentrated solution, nobelium was eluted just before Sr^{2+} tracer.

3.20 Some applications of the actinides

The use of uranium and plutonium in nuclear fuels and weapons has tended to overshadow the increasing number of useful (and peaceful) applications of isotopes of a number of these metals.

The oldest use, though not a current one worldwide, is that of thorium (as ThO_2) in incandescent gas mantles from the 1880s. Despite the increased use of electricity, such mantles are still widely used in the USA for gasolene lanterns for cooking as well as in decorative gas lights. The known catalytic potential of ThO_2 has thus far been little explored.

One side effect of the short half-lives and intense radioactivity of many actinides is the heat produced, and ^{238}Pu has found particular applicability in this respect (1 g of ^{238}Pu produces about 0.56 W). The Apollo-12 Lunar Mission left behind a generator powered by a 3.7 kg $^{238}PuO_2$ generating just under 1.5 kW of heat converted into electricity by thermoelectric elements, and similar power sources have been used on orbiting satellites as well as missions to photograph Jupiter and other outposts of the Solar System. More widely-used, however, is the application of ^{238}Pu to power heart pacemakers, where the use of nuclear power source lengthens the period between replacement by at least a factor of 5 compared with that when the pacemakers are powered by conventional chemical batteries. Such devices use around 160 mg of ^{238}Pu, encased in a Ta–Ir–Pt alloy; the heat output is converted to a regular electrical pulse which stimulates the heart muscle. ^{242}Cm has a higher heat output (122 Wg^{-1}) but is less widely used than ^{238}Pu, which is obtainable on the kilogram scale.

Another important range of applications involves neutron sources. Many of these use ^{241}Am as an α-particle source: on colliding with beryllium, neutrons are generated:

$$^{9}_{4}Be + ^{4}_{2}He \longrightarrow ^{12}_{6}C + ^{1}_{0}n$$

^{252}Cf produces neutrons directly as a result of spontaneous fission. Such small neutron sources are particularly convenient when nuclear reactors are not accessible; they find application in neutron activation analysis and neutron radiography. ^{241}Am sources are used as ionising sources for smoke detectors. One particular application of ^{252}Cf lies in plasma desorption mass spectroscopy (PDMS); when the high-energy fission fragments hit thin films of high molecular-weight molecules, they cause desorption of ions of very high mass from the film. This has been applied to the study of molecular ions with masses in excess of 50 000 from proteins; there is obvious utility to the study of non-volatile molecules of high molecular mass, which are not amenable to study by conventional mass spectroscopy.

3.21 Actinide organometallic chemistry

Although largely confined to uranium and thorium, this area has expanded rapidly over the last 15 years (as for the lanthanides); little study has been made of the other actinides, as problems arising from the air-sensitivity of organometallics are multiplied by the precautions necessary for the safe handling of the more radioactive (and in the case of the Pu, highly toxic) metals. So far as is known, the properties of corresponding compounds of different actinides in the same oxidation state are similar.

Most of the organometallic chemistry is that of the 4+ oxidation state, but there are a number of M(III) compounds.

Oxidation state (III)

$U(C_5H_5)_3$ is a typical compound, resembling the corresponding essentially ionic lanthanide(III) cyclopentadienyls:

$$UCl_3 + 3NaC_5H_5 \longrightarrow U(C_5H_5)_3 + 3NaCl$$

It readily forms quasi-tetrahedral adducts $U(C_5H_5)_3.L$ with a variety of Lewis bases (L = THF, $C_6H_{11}NC$ etc.) in which the cyclopentadienyl rings are *pentahapto*.

Analogous compounds $M(C_5H_5)_3$ are known for other actinides (M = Th, Pu, Am, Bk, Cm, Cf); the rare Th(III) compound should be noted. Similar tris(indenyls) are known for thorium and uranium, but use

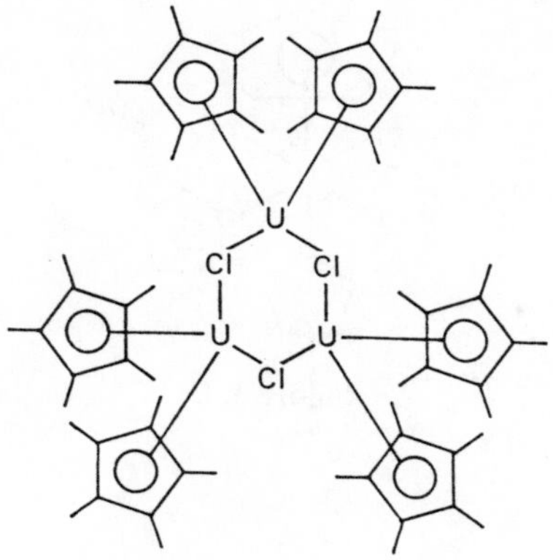

Figure 3.34

of the bulky pentamethylcyclopentadienyl ligand affords the green crystalline $U(C_5Me_5)_2Cl$, which has a trimeric solid-state structure (Figure 3.34). It can be turned into other U(III) compounds by substitution of Cl or by adduct formation with Lewis bases, to afford: $(C_5Me_5)_2UX$ (X = $CH(SiMe_3)_2$, $N(SiMe_3)_2$) and $(C_5Me_5)_2UCl.L$ (L = py, PMe_3, THF).

Oxidation state (IV) – cyclopentadienyls

Three series of compounds exist: $M(C_5H_5)_4$ (M = Th, Pa, U, Np); $M(C_5H_5)_3Cl$ (M = U, Th, Np); and $MCl_3(C_5H_5)$ (M = Th, U), prepared thus:

$$UCl_4 + 4KC_5H_5 \xrightarrow{C_6H_6} U(C_5H_5)_4 + 4KCl$$

$$ThCl_4 + 3TlC_5H_5 \xrightarrow{DEME} Th(C_5H_5)_3Cl + 3TlCl$$

$$UCl_4 + TlC_5H_5 \xrightarrow{THF} U(C_5H_5)Cl_3(THF)_2 + TlCl$$

The first two series have neo-tetrahedral coordination of the metal, while the adducts of $M(C_5H_5)Cl_3$ are pseudo-octahedral (Figure 3.35).

Chlorine can be replaced by other groups, thus $M(C_5H_5)_3X$ (X = NCS, BH_4, acac, CH_3, C_6H_5, for example) exist. The bonding in these M(IV) cyclopentadienyls is believed to have a definite covalent component; in contrast to $M(C_5H_5)_3$, $U(C_5H_5)_3Cl$ does not react with $FeCl_2$ to form ferrocene. In all these compounds, the C_5H_5 ring is bound *pentahapto*. Similar compounds have been made with the indenyl group which binds through the 5-membered ring:

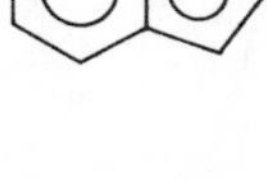

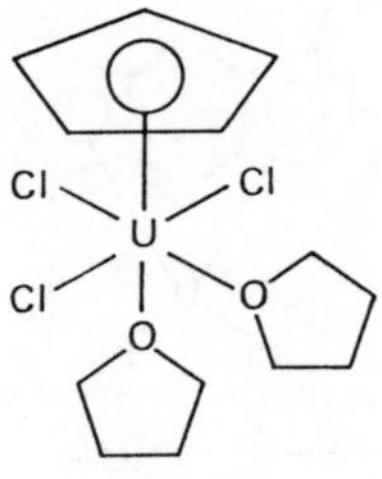

Figure 3.35

$U(C_5H_5)_2Cl_2$ does not exist but similar compounds can be made:

$$2C_5H_6 + U(NEt_2)_4 \longrightarrow U(C_5H_5)_2(NEt_2)_2$$

$$UCl_4 \xrightarrow{NaBH_4} UCl_2(BH_4)_2 \xrightarrow{TlC_5H_5} U(C_5H_5)_2(BH_4)_2$$

Using the pentamethylcyclopentadienyl ligand has a more dramatic effect; use of the ligand improves both crystallisability and solubility of the complex in hydrocarbons, but its steric bulk prevents more than two rings bonding to the metal:

$$MCl_4 + 2C_5Me_5MgBr \longrightarrow 2MgClBr + M(C_5Me_5)_2Cl_2$$

(M = Th, U)

These compounds have pseudo-tetrahedral structures (Figure 3.36).

Again, the chloride ligands can be replaced by other groups to afford $M(C_5Me_5)_2XY$ (X and/or Y = Cl, alkyl, OR, NR_2). The alkylamides have interesting carbonylation (CO insertion) reactions:

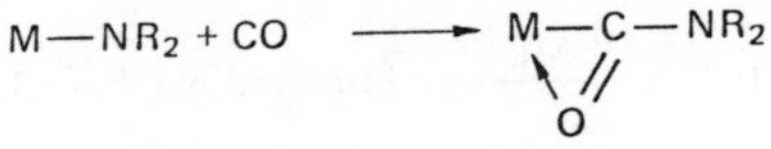

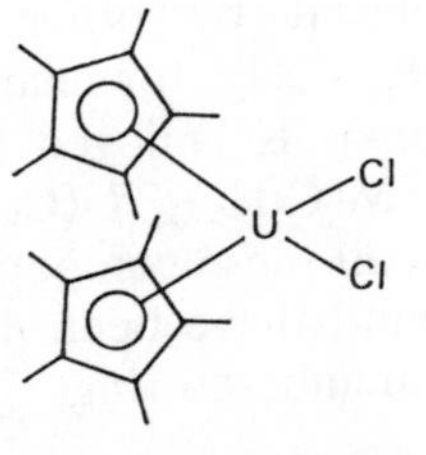

$$M{-}NR_2 + CO \longrightarrow M{-}C(=O){-}NR_2 \text{ (O also bonded to M)}$$

Figure 3.36

Cyclooctatetraenyl compounds

Using the cyclooctatetraenyl ligand, $C_8H_8^{2-}$, affords $U(C_8H_8)_2$, 'uranocene':

$$UCl_4 + 2K_2C_8H_8 \longrightarrow 4KCl + U(C_8H_8)_2$$

This has the 'sandwich' structure shown in Figure 3.37. Although the green solid is pyrophoric in air, it is chemically unreactive in other ways, undergoing only slow hydrolysis. Similar (but more reactive) compounds have been prepared for thorium, protactinium, neptunium and plutonium.

The U–C_8H_8 bond dissociation energy in uranocene has been measured at 347 kJ mol^{-1}, significantly greater than the Fe–C_5H_5 bond energy in ferrocene (297 kJ mol^{-1}). Overlap of filled ligand MOs with uranium 5*f*

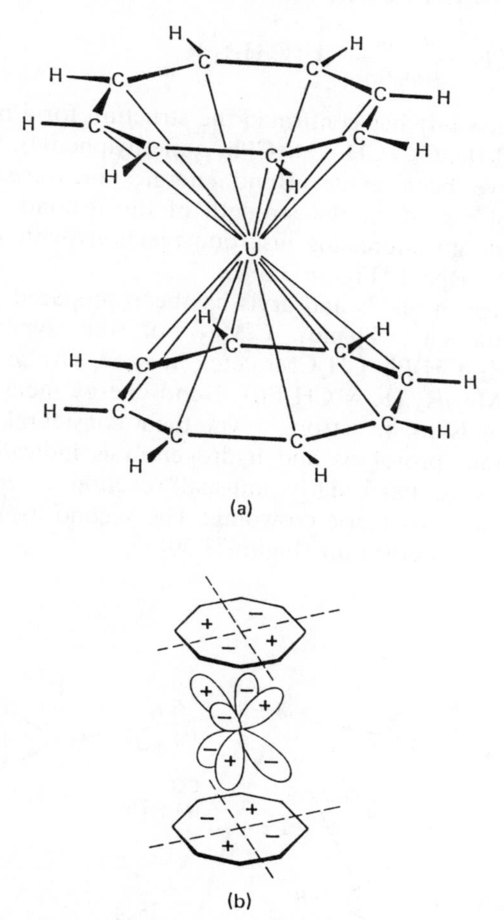

Figure 3.37 $(C_8H_8)_2U$, uranocene: (a) structure; (b) metal–ligand interaction involving *f* orbitals

orbitals has been invoked to explain the bonding; it is likely that 6*d* orbitals are just as (if not more) important.

A number of ring-substituted uranocenes have been synthesised; of these, $U(C_8H_4Ph_4)_2$ is air-stable, doubtless owing to the use of bulky ligands that restrict access to the uranium atom. 'Half-sandwich' compounds like $Th(C_8H_8)Cl_2(THF)_2$ have also been made.

σ-bonded organometallics

Early attempts to make UR_n species (R = alkyl or aryl) were unsuccessful. In the early 1970s, the first reports were made of the compounds $U(C_5H_5)_3R$ (R = CH_3, C_6H_5, C_4H_9, CMe_3, C≡CPh, for example):

$$U(C_5H_5)_3Cl \xrightarrow[\text{RMgBr}]{\text{RLi or}} U(C_5H_5)_3\,R$$

Crystallographic study has confirmed the structure for a number of these ligands (R = C_4H_9, C≡CH, C≡CPh, *p*-methylphenyl). Similar thorium compounds have been made. Bond energies measured in the range 320–360 kJ mol^{-1} confirm the strength of the σ-bond. Some of these compounds undergo interesting insertion reactions with small molecules (CO, CO_2, SO_2, RNC) (Figure 3.38).

A wide range of alkyls and aryls has been prepared with the pentamethylcyclopentadienyl ligand, C_5Me_5, of the type $M(C_5Me_5)_2R_2$ (R = CH_3, C_6H_5, CH_2Ph, CH_2CMe_3 etc.; M = U, Th), as well as a few of the type $M(C_5Me_5)R_3$ (R = CH_2Ph). Bond energy measurements again indicate the bonds to be 'strong' – yet their considerable readiness to undergo insertion, protolysis and hydrogenolysis indicates a significant bond polarity. One particularly unusual reaction is cyclometallation, apparently favoured by steric crowding. The second example illustrated shows a notable isomerisation (Figure 3.39).

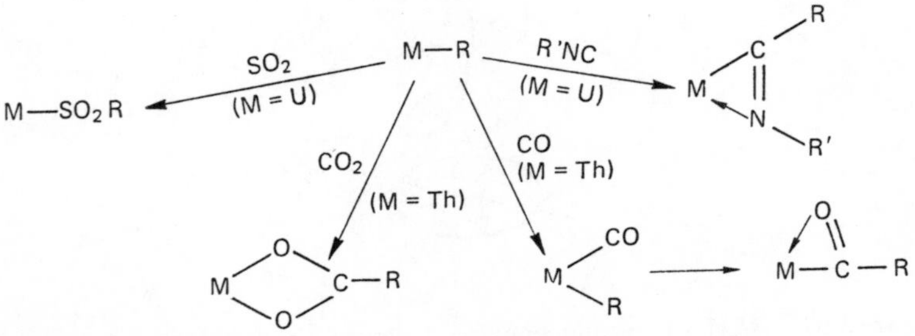

Figure 3.38

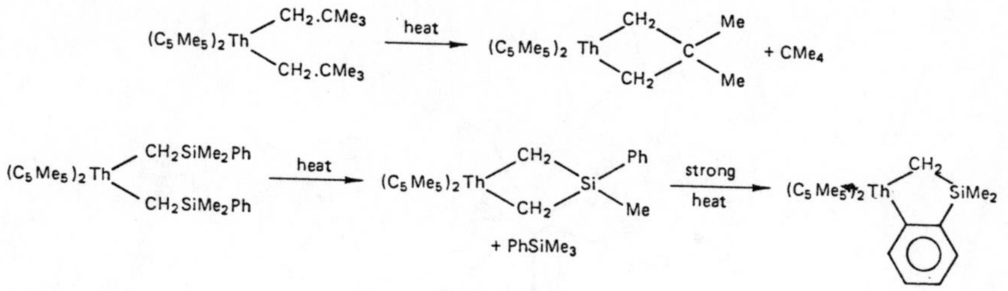

Figure 3.39

Hydrogenolysis of the alkyls affords remarkable hydrides:

$$M(C_5Me_5)_2R_2 + 4H_2 \longrightarrow M(C_5Me_5)_2H_2 + 4RH$$

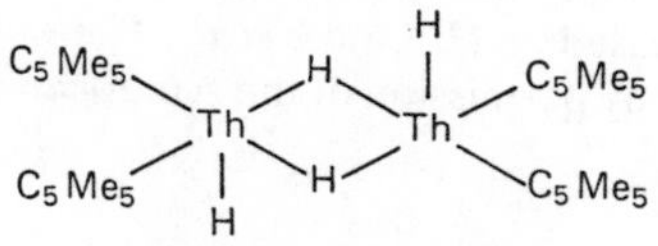

Figure 3.40

The thorium compound has been shown to have a dimeric structure (Figure 3.40) with both bridging and terminal hydrogens. In solution, NMR shows rapid exchange, not just between bridging and terminal hydrogens, but also with dissolved hydrogen gas. This compound and the uranium analogue are extremely efficient catalysts for α-alkene hydrogenation and ethene polymerisation.

Homoleptic σ-bonded compounds

Since the failure of early attempts to prepare UR_n (n = 4–6, R = alkyl or aryl), successful methods have utilised three routes. Stabilisation via coordinative saturation in anionic complexes has permitted the isolation of UR_6^{2-}, UR_8^{3-} and $ThMe_7^{3-}$ as Li^+ salts. The thorium compound seems to owe part of its stability to Li ... H–C interactions (Figure 3.41).

A second approach involves suitable 'supporting' ligands, in this case chelating tertiary phosphines:

$$MCl_4(dmpe)_2 + 4RLi \longrightarrow 4LiCl + MR_4(dmpe)_2$$

$$(R = CH_3, CH_2Ph;\ M = Th, U;\ dmpe = Me_2P(CH_2)_2PMe_2)$$

The phosphines may stabilise the complex largely by occupying coordination sites, though there is evidence for *polyhapto* Th–benzyl interaction. Tetrabenzylthorium itself is a stable, colourless, crystalline solid of as yet unknown structure, where extra stabilisation could be conferred by metal–ring interaction (Figure 3.42).

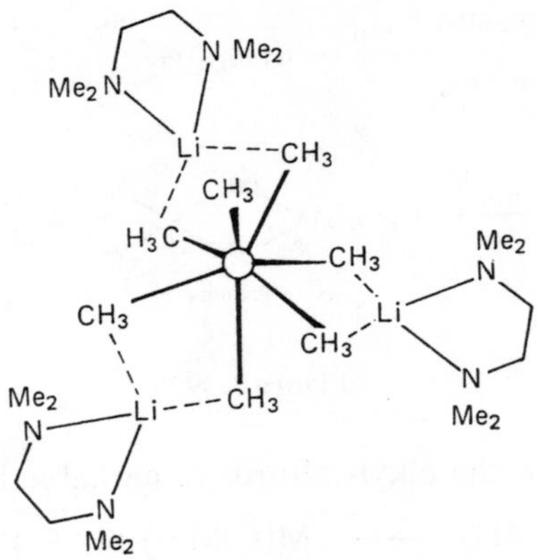

Figure 3.41 $ThMe_7^{3-}$ ions stabilised in the $Li(TMEDA)^+$ complex. Note the Li^+–CH_3 interaction [based on H. Lauke *et al.*, *J. Amer. Chem. Soc.*, **106** (1984) 6841; reprinted by permission of the American Chemical Society]

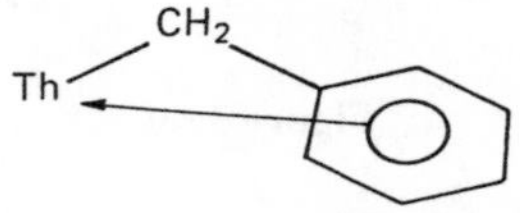

Figure 3.42

Thirdly, the use of the correct ligands seems to be important, as in the first homoleptic uranium alkyl recently reported:

$$U(OR)_3 \xrightarrow{LiCH(SiMe_3)_2} U[CH(SiMe_3)_2]_3$$

(R = 2,6-di-t-butylphenyl)

The ligand is similar in steric requirements to the amide $N(SiMe_3)_2$, which similarly affords stable three coordinate compounds.

It is characteristic of the transition metals that they form stable binary carbonyls. Although the early actinides show some resemblance to the early *d*-block metals, uranium carbonyls such as $U(CO)_6$ are only stable in matrices up to 30K. The compound $[Me_3SiC_5H_4]_3U.CO$ exists, but cannot be isolated in the solid state.

Metal–metal bonded organometallics

A number of compounds like $U[Mn(CO)_5]_4$ containing bridging carbonyls have been known for some years, but it is only recently that the first actinide–transition-metal bonds were definitely characterised, via a conventional salt-elimination reaction:

$$Th(C_5Me_5)_2X_2 + (C_5H_5)Ru(CO)_2Na \xrightarrow{\text{toluene}} NaX + (C_5Me_5)_2Th{-}Ru(C_5H_5)(CO){-}CO$$

$$(X = Cl, I)$$

3.22 Synthesis and properties of elements 104–109

Over the last quarter century, numerous attempts have been made to synthesise elements with atomic numbers greater than 103, expected to form a '6*d*' transition series. Although the trend towards shorter half-life with increasing atomic number, established in the actinide series, continues, it has been predicted that eventually nuclei will become stabler as the atomic number approaches 114. Though this goal has not yet been achieved, there is evidence for the synthesis of elements 104–109.

Element 104 (rutherfordium/kurchatovium)

At present there are conflicting claims from researchers in the USA and the USSR, which means that the name for the element has not been approved by IUPAC. Nuclei of masses 253–262 have been claimed, and two seem to be particularly well-characterised:

$$^{208}Pb + {}^{58}Ti \longrightarrow {}^{257}104 + {}^{1}_{0}n$$

$$^{248}Cm + {}^{18}O \longrightarrow {}^{261}104 + 5{}^{1}_{0}n$$

The latter isotope has a half-life of 1.1 m and is an α-emitter, unlike many heavy nuclei, which tend to decompose by spontaneous fission.

Some elegant (and rapid) chemistry has been carried out on element 104. Thus in a series of several hundred repeated experiments, US workers synthesised about 100 atoms of $^{261}104$, studied their elution behaviour on a cation-exchange resin, compared it with Zr^{4+}, Hf^{4+}, No^{2+} and actinide^{3+}, and found it to be very similar to zirconium and hafnium. Complexing studies with chloride support this view. In a third series of experiments, workers in the USSR produced atoms now believed to be $^{259}104$ thus:

$$^{242}Pu + {}^{22}Ne \longrightarrow {}^{259}104 + 5{}^{1}_{0}n$$

The atoms were collected in a gas stream, chlorinated with $SOCl_2$ and passed through a heated 2-metre glass column. The retention times in the column were found to be similar to those for $ZrCl_4$ and $HfCl_4$, but significantly different from those for actinide chlorides.

The overall view is thus that element 104 is indeed eka-hafnium.

Element 105 ('hahnium')

Various isotopes have been reported and conflicting USA/USSR claims have again complicated matters. Syntheses include:

$$^{243}\text{Am} + {}^{22}\text{Ne} \longrightarrow {}^{261}105 + 4{}^{1}_{0}\text{n}$$

$$^{249}\text{Bk} + {}^{18}\text{O} \longrightarrow {}^{262}105 + 5{}^{1}_{0}\text{n}$$

These isotopes have half-lives of 1.8 s and 34 s respectively. Gas-phase studies (on atoms with a 2-second half-life!) suggest that element 105 forms a chloride less volatile than $NbCl_5$ but more volatile than $HfCl_4$, as expected for eka-tantalum in Group 5 (VB).

Elements 106–109

Four isotopes of elements 106 have been claimed, the most stable being $^{263}106$:

$$^{249}\text{Cf} + {}^{18}\text{O} \longrightarrow 4{}^{1}_{0}\text{n} + {}^{263}106\ (\alpha,\ 0.9\ \text{s})$$

It might be expected to form a volatile fluoride similar to WF_6, but no chemical studies have yet been reported.

For elements 107–109, only very short-lived isotopes have yet been synthesised:

$$^{209}\text{Bi} + {}^{54}\text{Cr} \longrightarrow {}^{1}_{0}\text{n} + {}^{262}107\ (\alpha,\ 4.7\ \text{ms})$$

$$^{208}\text{Pb} + {}^{58}\text{Fe} \longrightarrow {}^{1}_{0}\text{n} + {}^{265}108\ (\alpha,\ 1.8\ \text{ms})$$

$$^{209}\text{Bi} + {}^{58}\text{Fe} \longrightarrow {}^{1}_{0}\text{n} + {}^{266}109\ (\alpha,\ 3.5\ \text{ms})$$

They would be expected to be analogues of Re, Os, and Ir respectively, but again, no chemical properties are known.

3.23 Superheavy elements?

Synthesis of the transuranium elements, particularly those with atomic numbers beyond 100, has become progressively more difficult. Half-lives of the isotopes are smaller, and there is an increasing tendency to spontaneous fission.

Theoretical study has predicted the existence of 'superheavy' elements with atomic numbers in the region of 110–120 whose nuclei will be more stable than those of the preceding elements. Modern views of nuclear structure ascribe particular stability to filled shells of protons and neutrons; thus ^{208}Pb with 82 protons and 126 neutrons is regarded as especially stable as both the protons and neutrons correspond to filled subshells – 'magic' numbers. An element with 114 protons and 184 neutrons would again have filled subshells, and interest has centred on synthesising it, and its neighbours in the Periodic Table. Attempts have included:

$$^{248}Cm + {}^{48}Ca \not\rightarrow {}^{295}116 + {}^{1}_{0}n$$

Syntheses so far have been unsuccessful, but further attempts will doubtless continue.

Appendix A: Ionisation energies

Lanthanide ionisation energies (kJ mol^{-1})

	I_1	I_2	I_3	I_4
Sc	631	1235	2389	7089
Y	616	1181	1980	5963
La	538.1	1067	1850	4819
Ce	527.4	1047	1949	3547
Pr	523.1	1018	2086	3761
Nd	529.6	1035	2130	3899
Pm	523.9	1052	2150	3970
Sm	543.3	1068	2260	3990
Eu	546.7	1085	2404	4110
Gd	592.5	1167	1990	4250
Tb	564.6	1112	2114	3839
Dy	571.9	1126	2200	4501
Ho	580.7	1139	2204	4150
Er	588.7	1151	2194	4115
Tm	596.7	1163	2285	4119
Yb	603.4	1176	2415	4220
Lu	523.5	1340	2022	4360

Data from W. C. Martin *et al.*, *J. Phys. Chem. Ref. Data*, **3** (1974) 771.

Actinide ionisation energies (kJ mol^{-1})

	I_1	I_2	I_3	I_4
Ac	499	1170	1900	(4700)
Th	587	1110	1978	2780
Pa	568	1128	—	2991
U	584	1420	1900	3145
Np	597	1128	1997	3242
Pu	585	1128	2084	3338
Am	578	1158	2132	3493
Cm	581	1196	2026	3550
Bk	601	1186	2152	3434
Cf	608	1206	2267	3599
Es	620	1216	2334	3734
Fm	627	1225	2363	3792
Md	635	1235	2470	3840
No	642	1254	2643	3956
Lr	444	1428	2228	4910

Appendix B: Ionic radii

Lanthanide and actinide ionic radii (Å)

	CN 6	CN 8		CN 6	CN 8
Sc^{3+}	0.745	0.870			
Y^{3+}	0.900	1.019			
La^{3+}	1.032	1.160	Ac^{3+}	1.12	
Ce^{3+}	1.01	1.143	Th^{3+}	—	
Pr^{3+}	0.99	1.126	Pa^{3+}	1.04	
Nd^{3+}	0.983	1.109	U^{3+}	1.025	
Pm^{3+}	0.97	1.093	Np^{3+}	1.01	
Sm^{3+}	0.958	1.079	Pu^{3+}	1.00	
Eu^{3+}	0.947	1.066	Am^{3+}	0.975	1.09
Gd^{3+}	0.938	1.053	Cm^{3+}	0.97	
Tb^{3+}	0.923	1.040	Bk^{3+}	0.96	
Dy^{3+}	0.912	1.027	Cf^{3+}	0.95	
Ho^{3+}	0.901	1.015			
Er^{3+}	0.890	1.004			
Tm^{3+}	0.880	0.994			
Yb^{3+}	0.868	0.985			
Lu^{3+}	0.861	0.977			
Nd^{2+}	—	1.29			
Sm^{2+}	—	1.27			
Eu^{2+}	1.17	1.25			
Dy^{2+}	1.07	1.19			
Tm^{2+}	1.03	—			
Yb^{2+}	1.02	1.14			
Ce^{4+}	0.87	0.97	Th^{4+}	0.94	1.05
Pr^{4+}	0.85	0.96	Pa^{4+}	0.90	1.01
Tb^{4+}	0.76	0.88	U^{4+}	0.89	1.00
			Np^{4+}	0.87	0.98
			Pu^{4+}	0.86	0.96
			Am^{4+}	0.85	0.95
			Cm^{4+}	0.85	0.95
			Bk^{4+}	0.83	0.93
			Cf^{4+}	0.821	0.92
			Pa^{5+}	0.78	0.91
			U^{5+}	0.76	—
			Np^{5+}	0.75	—
			Pu^{5+}	0.74	—
			U^{6+}	0.73	0.86
			Np^{6+}	0.72	—
			Pu^{6+}	0.71	—

Data from R. D. Shannon, *Acta Crystallographica A*, **32** (1976) 752.

Appendix C: Periodic Table

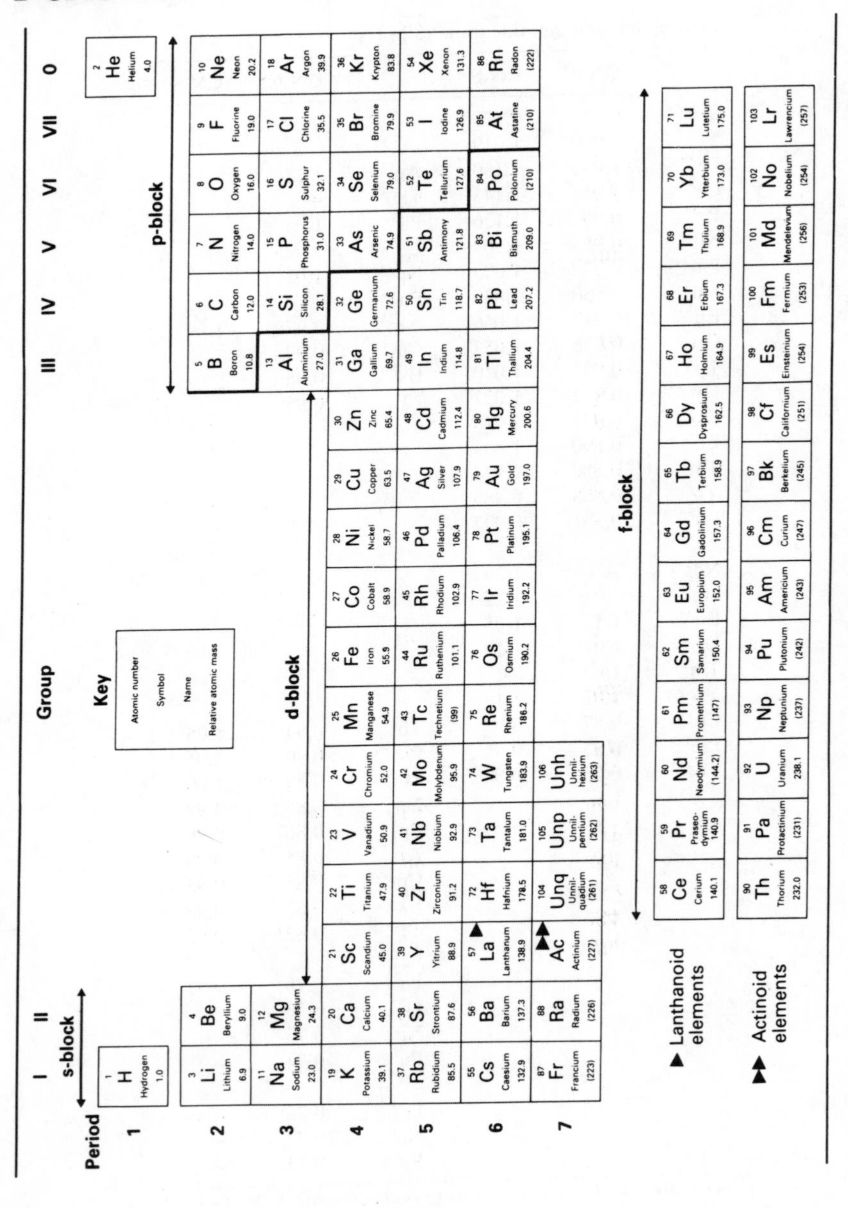

Group

Key

Atomic number
Symbol
Name
Relative atomic mass

s-block · d-block · p-block · f-block

Period	I	II											III	IV	V	VI	VII	0
1	1 H Hydrogen 1.0																	2 He Helium 4.0
2	3 Li Lithium 6.9	4 Be Beryllium 9.0											5 B Boron 10.8	6 C Carbon 12.0	7 N Nitrogen 14.0	8 O Oxygen 16.0	9 F Fluorine 19.0	10 Ne Neon 20.2
3	11 Na Sodium 23.0	12 Mg Magnesium 24.3											13 Al Aluminium 27.0	14 Si Silicon 28.1	15 P Phosphorus 31.0	16 S Sulphur 32.1	17 Cl Chlorine 35.5	18 Ar Argon 39.9
4	19 K Potassium 39.1	20 Ca Calcium 40.1	21 Sc Scandium 45.0	22 Ti Titanium 47.9	23 V Vanadium 50.9	24 Cr Chromium 52.0	25 Mn Manganese 54.9	26 Fe Iron 55.9	27 Co Cobalt 58.9	28 Ni Nickel 58.7	29 Cu Copper 63.5	30 Zn Zinc 65.4	31 Ga Gallium 69.7	32 Ge Germanium 72.6	33 As Arsenic 74.9	34 Se Selenium 79.0	35 Br Bromine 79.9	36 Kr Krypton 83.8
5	37 Rb Rubidium 85.5	38 Sr Strontium 87.6	39 Y Yttrium 88.9	40 Zr Zirconium 91.2	41 Nb Niobium 92.9	42 Mo Molybdenum 95.9	43 Tc Technetium (99)	44 Ru Ruthenium 101.1	45 Rh Rhodium 102.9	46 Pd Palladium 106.4	47 Ag Silver 107.9	48 Cd Cadmium 112.4	49 In Indium 114.8	50 Sn Tin 118.7	51 Sb Antimony 121.8	52 Te Tellurium 127.6	53 I Iodine 126.9	54 Xe Xenon 131.3
6	55 Cs Caesium 132.9	56 Ba Barium 137.3	57 ▶ La Lanthanum 138.9	72 Hf Hafnium 178.5	73 Ta Tantalum 181.0	74 W Tungsten 183.9	75 Re Rhenium 186.2	76 Os Osmium 190.2	77 Ir Iridium 192.2	78 Pt Platinum 195.1	79 Au Gold 197.0	80 Hg Mercury 200.6	81 Tl Thallium 204.4	82 Pb Lead 207.2	83 Bi Bismuth 209.0	84 Po Polonium (210)	85 At Astatine (210)	86 Rn Radon (222)
7	87 Fr Francium (223)	88 Ra Radium (226)	▶▶ Ac Actinium (227)	104 Unq Unnil-quadium (261)	105 Unp Unnil-pentium (262)	106 Unh Unnil-hexium (263)												

▶ Lanthanoid elements	58 Ce Cerium 140.1	59 Pr Praseo-dymium 140.9	60 Nd Neodymium (144.2)	61 Pm Promethium (147)	62 Sm Samarium 150.4	63 Eu Europium 152.0	64 Gd Gadolinium 157.3	65 Tb Terbium 158.9	66 Dy Dysprosium 162.5	67 Ho Holmium 164.9	68 Er Erbium 167.3	69 Tm Thulium 168.9	70 Yb Ytterbium 173.0	71 Lu Lutetium 175.0
▶▶ Actinoid elements	90 Th Thorium 232.0	91 Pa Protactinium (231)	92 U Uranium 238.1	93 Np Neptunium (237)	94 Pu Plutonium (242)	95 Am Americium (243)	96 Cm Curium (247)	97 Bk Berkelium (245)	98 Cf Californium (251)	99 Es Einsteinium (254)	100 Fm Fermium (253)	101 Md Mendelevium (256)	102 No Nobelium (254)	103 Lr Lawrencium (257)

Bibliography

Because of the increasing selectivity of libraries, where possible more than one source of further reading on a topic has been given.

For general properties of the elements, consult:

1. J. Emsley, *The Elements*, Oxford, 1989.
2. R. C. Weast (ed.), *CRC Handbook of Chemistry and Physics*, 70th edn, Central Rubber Company Press, 1989–90.

For a particular compound, three avenues are open:

1. A search via *Chemical Abstracts*.
2. Consult the appropriate volume of 'Gmelin'.
3. Consult either the *Dictionary of Organometallic Compounds* (Chapman and Hall, 1984, five supplements to 1989) or *Dictionary of Inorganic Compounds* (Chapman and Hall, due to be published in early 1992). Between them they contain data on well over 2000 lanthanide and actinide compounds.

General sources

Abbreviations used in text

ACE J. J. Katz, G. J. Seaborg and L. R. Morss (eds), *The Chemistry of the Actinide Elements*, 2nd edn, Chapman and Hall, 1986.

CCC G. Wilkinson, R. D. Gillard and J. A. McCleverty, (eds), *Comprehensive Coordination Chemistry*, Pergamon, 1987.

CIC J. C. Bailar Jr, H. J. Emeleus, R. S. Nyholm and A. F. Trotman-Dickenson (eds), *Comprehensive Inorganic Chemistry*, Pergamon, 1973.

LPL J-C. G. Bunzli and G. R. Choppin (eds), *Lanthanide Probes in Life, Chemical and Earth Sciences*, Elsevier, 1989.

HRE K. A. Gschneider Jr and LeRoy Eyring (eds), *Handbook on the Physics and Chemistry of Rare Earths*, North-Holland, 1978 onwards, vols 1–12.

HAC	A. J. Freeman and G. H. Lander (eds), *Handbook on the Physics and Chemistry of the Actinides*, North-Holland, 1984 onwards, vols 1–5.
LAS	N. M. Edelstein (ed), *Lanthanide and Actinide Chemistry and Spectroscopy*, ACS Symposium Series, 1980, vol. 131.
AIP	N. M. Edelstein (ed), *Actinides in Perspective*, Pergamon, 1982.
MTP1 *MTP2*	K. W. Bagnall (ed), *MTP International Review of Science, Inorganic Chemistry*, *Series* 1 (1972) vol. 7; *Series* 2 (1975) vol. 7.
SPL	S. P. Sinha (ed), *Systematics and Properties of the Lanthanides*, D. Reidel, 1983.
IAR	K. A. Gschneider Jr (ed.), *Industrial Application of Rare Earth Elements*, ACS Symposium Series, 1981.
GMEL	Gmelins *Handbuch der Anorganischer Chemie*.

Chapter 1 – Scandium

The Gmelin System covers scandium as a 'rare earth' in Syst. No. 39.
C. T. Horowitz (ed.), *Scandium*, Academic Press, 1975.
R. C. Vickery, *CIC*, **3**, 329 (general chemistry).
G. A. Melson *et al.*, *Coord. Chem. Revs*, **7** (1971) 133 (complexes).
F. A. Hart, *CCC*, **3**, 1059 (complexes).

Organometallics are generally reviewed with the lanthanides (see Chapter 2).

For the pentamethylcyclopentadienyls, see:
M. E. Thompson *et al.*, *Pure Appl. Chem.*, **56** (1984) 1; *J. Amer. Chem. Soc.*, **109** (1987) 203.

Chapter 2 – The lanthanides

For historical perspective and general coverage:
D. M. Yost, H. Russell Jr and C. S. Garner, *The Rare Earths and their Compounds*, John Wiley, 1947.
R. C. Vickery, *The Chemistry of Yttrium and Scandium*, Pergamon, 1960; *CIC*, **3**, 329.
T. Moeller, *CIC*, **4**, 1.
N. E. Topp, *The Chemistry of the Rare Earth Elements*, Elsevier, 1965.
G. R. Choppin, *LPL*, **1**.

Gmelin coverage of these elements in Series 39 is well up to date: Series A covers occurrence and minerals; Series B separation and metals; Series C binary compounds and salts; and Series D coordination compounds and organometallics.

Section 2.1

T. Moeller, *J. Chem. Educ.*, **47** (1970) 417 (principles).
W. B. Jensen, *J. Chem. Educ.*, **59** (1982) 634 (positions of La, Lu).
D. R. Lloyd, *J. Chem. Educ.*, **63** (1986) 502 (lanthanide contraction).
H. G. Friedman *et al.*, *J. Chem. Educ.*, **41** (1964) 354 (*f* orbitals).

Section 2.2

H. E. Kremers, *J. Chem. Educ.*, **62** (1985) 665 (extraction).
G. W. de Vore, *LPL*, 321 (geochemistry).
J.-C. Duchesne, *SPL*, 543 (geochemistry).
P. Moller, *SPL*, 561 (geochemistry).
S. R. Taylor and M. M. McLennan, *HRE*, **11**, 485 (geochemistry).
L. A. Haskin and T. P. Paster, *HRE*, **3**, 1 (geochemistry).

Section 2.3

B. J. Beaudry and K. A. Gschneider Jr, *HRE*, **1**, 173.
H. F. Linebarger and T. K. McCluhan; L. A. Luyckx; K. E. Davies, in *IAR*, 3; 19; 167 (applications).
E. L. Huston and J. J. Sheridan III, *IAR*, 223 (hydrides).
G. G. Libowitz and A. J. Maeland, *HRE*, **3**, 299 (hydrides).

Section 2.5

L. J. Nugent, *J. Inorg. Nucl. Chem.*, **37** (1975) 1767; *MTP2*, **7**, 195 (redox potentials, oxidation states).
D. A. Johnson, *Some Thermodynamic Aspects of Inorganic Chemistry*, 2nd edn, Cambridge U.P., 1982, 50–54, 158–168; *J. Chem. Educ.*, **57** (1980) 475.
D. W. Smith, *Inorganic Substances*, Cambridge U.P., 1990, 145–149, 163–167; *J. Chem. Educ.*, **63** (1986) 228.

Section 2.6

Electronic spectra

F. A. Hart, *CCC*, **3**, 1105.
T. Moeller, *CIC*, **4**, 9.
K. B. Yatsimirskii and N. H. Davidenko, *Coord. Chem. Revs*, **27** (1979) 223.
S. Hufner, *SPL*, 313.
W. T. Carnall *et al.*, *SPL*, 389; *HRE*, **3**, 171.
C. A. Morrison and R. P. Leavitt, *HRE*, **5**, 461.
J. W. O'Laughlin, *HRE*, **4**, 341.
B. R. Judd, *HRE*, **11**, 81.

Fluorescence

R. E. Whan and G. A. Crosby, *J. Mol. Spectrosc.*, **8** (1962) 315.
F. S. Richardson, *Chem. Revs*, **82** (1982) 541.
J-C. G. Bunzli, *LPL*, 219.
G. Blasse, *HRE*, **4**, 237.

Colour TV

J. R. McColl and F. C. Palilla, *IAR*, 177.
G. Blasse, *HRE*, **4**, 237.

Lasers

M. J. Weber, *HRE*, **4**, 275.

Magnetism

F. A. Hart, *CCC*, **3**, 1109.
T. Moeller, *CIC*, **4**, 9.
J. Rossat-Mignod, *SPL*, 255.

Section 2.7 (see also Section 2.6)

T. Moeller, *MTP1*, **7**, 275 (rev.).
J. H. Forsberg, *Coord. Chem. Revs*, **10** (1973) 195 (rev.).
D. G. Karraker, *Inorg. Chem.*, **7** (1968) 473 (hypersens.).
B. R. Judd, *LAS*, 267.
J. L. Ryan and C. K. Jorgensen, *J. Phys. Chem.*, **70** (1966) 2845.
E. M. Stephens, *LPL*, 181 (Gd ESR).
A. Abragam and B. Bleaney, *Electron Paramagnetic Resonance of Transition Ions*, Oxford, 1970 (Ln ESR).
J. Reuben and G. A. Elgarish, *HRE*, **4**, 483 (NMR).
F. A. Hart, *CCC*, **3**, 1100 (NMR).
J-C. G. Bunzli, *LPL*, 219 (bioprobes).
W de W. Horrocks Jr and D. R. Sudnick, *Acc. Chem. Res.*, **14** (1981) 384 (bioprobes).
J. Reuben, *HRE*, **4**, 515 (bioprobes).
J. R. Ascenso and A. V. Xavier, *SPL*, 501 (bioprobes).
F. S. Richardson, *Chem. Revs*, **82** (1982) 541 (bioprobes).

Section 2.8

D. Brown, *Halides of the Lanthanides and Actinides*, John Wiley, 1968.
J. Burgess and J. Kijorski, *Adv. Inorg. Chem. Radiochem.*, **24** (1981) 51 (halides).
J. M. Haschke, *HRE*, **4**, 89 (halides).
L. Eyring, *HRE*, **3**, 337 (oxides).
GMEL: *C1* (oxides), *C2* (hydroxides), *C3*, *C4a/b*, *C6* (halides), *C7* (sulphides), *C11a* (borides).
P. P. Edwards *et al.*, *Chem. Brit.*, **23** (1987) 962 (superconductivity).

K. A. Bednorz and J. G. Muller, *Science*, **237** (1987) 1133; *Angew. Chem. Int. Ed.*, **27** (1988) 735 (superconductivity).
C. N. R. Rao (ed.), *Chemistry of Oxide Superconductors*, Blackwell, Oxford, 1988.

Section 2.9

T. Moeller, *CIC*, **4** (1973) 28; *MTP1*, 285 (stability constants).
A. E. Martell and R. M. Smith, *Critical Stability Constants*, Plenum, 1974 onwards, vols 1–6.

Section 2.10

GMEL, D1–D5.
F. A. Hart, *CCC*, **3**, 1059.
L. Niinsto and M. Leskala, *HRE*, **8**, 203; **9**, 91 (inorganic complexes).
L. C. Thompson, *HRE*, **3**, 209.
J. F. Desreux, *LPL*, 43.
J-C. G. Bunzli, *HRE*, **9**, 321; *LPL*, 219 (macrocycles).

Section 2.11

Reviews on shift reagents include:

J. Reuben, *Prog. Nucl. Magn. Res. Spectrosc.*, **9** (1975) 1.
J. Reuben and G. A. Elgarish, *HRE*, **4**, 483.
A. D. Sherry and C.F.G.C. Geraldes, *LPL*, 219.
M. R. Peterson Jr and G. H. Wahl Jr, *J. Chem. Educ.*, **49** (1972) 790.
R. E. Sievers, *Nuclear Magnetic Shift Reagents*, Academic Press, 1973.
F. A. Hart, *CCC*, **3**, 1105.

M. F. Tweedle, *LPL*, 127 (NMR imaging).
R. B. Lauffer, *Chem. Rev.*, **87** (1987) 901 (NMR imaging).

Section 2.12

D. A. Johnson, *Adv. Inorg. Chem. Radiochem.*, **20** (1977) 1.
F. A. Hart, *CCC*, **3**, 1109, 1113.
T. Moeller, *CIC*, **4**, 75, 97.

Section 2.13

F. Weigel, *Chem. Zeitung*, **108** (1978) 339.
G. E. Boyd, *J. Chem. Educ.*, **36** (1959) 3.
E. J. Wheelwright (ed.), *Promethium Technology*, American Nuclear Society, 1973.

W. R. Wilmarth *et al.*, *J. Less Common Metals*, **141** (1988) 275; *J. Raman Spectroscopy*, **19** (1988) 271.

Section 2.14

T. J. Marks and R. D. Fischer (eds), *Organometallic Chemistry of the f-block Elements*, Reidel, 1979.
W. J. Evans, *Adv. Organomet. Chem.*, **24** (1985) 131.
H. Schumann and W. Genthe, *HRE*, **7**, 440.
T. J. Marks, *Progr. Inorg. Chem.*, **24** (1978) 51.
S. A. Cotton, *J. Organomet. Chem. Library*, **3** (1977) 217.
P. L. Watson and G. W. Parshall, *Acc. Chem. Res.*, **88** (1985) 51 (catalysis).
G. Jeske *et al.*, *J. Amer. Chem. Soc.*, **107** (1985) 8111 (catalysis).
H. B. Kagan and J. L. Namy, *HRE*, **6**, 523 (organic synthesis).
J. R. Long, *HRE*, **8**, 235; *Inorg. Chim. Acta*, **18** (1985) 87 (organic synthesis).
T. J. Marks and R. D. Ernst, *Comprehensive Organometallic Chemistry*, Pergamon, **3** (1982) 173.

Chapter 3 – The actinides

Apart from the text by Katz, Seaborg and Morss, general (though less up to date) coverage is given by:

C. Keller, *The Chemistry of the Transuranium Elements*, Verlag Chemie, 1971.
K. W. Bagnall, *The Actinide Elements*, Elsevier, 1972.

Section 3.2

J. J. Katz, G. T. Seaborg and L. R. Morss, *ACE*, **2**, 1122.
C. Keller, *Angew. Chem. Int. Ed.*, **4** (1965) 903.
G. T. Seaborg, *Man-made Transuranium Elements*, Prentice-Hall, 1963.

Complete lists of isotopes with half lives can be found in, for example, *ACE*, pp. 1654–1668; *CRC Handbook of Chemistry and Physics* (Rubber Handbook), B228–448 (1989 edition).

Section 3.3

M. S. Fred, *ACE*, **2**, 1196.
J. J. Katz, L. R. Morss and G. T. Seaborg, *ACE*, **2**, 1133, 1142.
L. J. Nugent, *J. Inorg. Nucl. Chem.*, **37** (1975) 1767; *MTP2*, **7**, 195.
K. W. Bagnall, *Essays in Chemistry*, **3** (1972) 39.

Section 3.5

L. I. Katzin and D. C. Sonnenberger, *ACE*, **1**, 46 (thorium).
Z. Kolaria, *HAC*, **3**, 431.
Z. Yong-jun, *HAC*, **3**, 469.
F. Weigel, *ACE*, **1**, 196–211 (uranium).
H. W. Kirby, *ACE*, **1**, 112–118 (protactinium).

Section 3.6

F. Weigel, *ACE*, **1**, 173.
S. Villani, *Isotope Separation*, American Nuclear Society, 1976; *Uranium Enrichment*, Springer-Verlag, 1979.

Section 3.7

B. Allard *et al.*, *Inorg. Chim. Acta*, **94** (1984) 205.
K. N. Raymond *et al.*, *AIP*, 491; *Inorg. Chim. Acta*, **94** (1984) 193; *Inorg. Chem.*, **25** (1985) 605.
R. A. Belman, *Struct. Bonding*, **34** (1978) 39.
J. J. Katz, G. T. Seaborg and L. R. Morss, *ACE*, **2**, 1169.
J. R. Duffield and D. M. Taylor, *HAC*, **4**, 129.
G. R. Choppin and B. Allard, *HAC*, **3**, 407.

Section 3.8

E. K. Hulet and D. D. Bode, *MTPI*, **7**, 1.
F. Weigel, J. J. Katz and G. T. Seaborg, *ACE*, **1**, 505–577.

Section 3.9

E. K. Hulet *et al.*, *J. Inorg. Nucl. Chem.*, **42** (1980) 79.

Section 3.10

D. Brown, *Halides of Lanthanides and Actinides*, John Wiley, 1968; *MTP*, **1**, 87.
J. C. Taylor, *Coord. Chem. Revs*, **20** (1976) 197.

Section 3.11

J. L. Ryan, *MTP1*, **7**, 323 (UV–visible).
W. T. Carnall and H. M. Crosswhite, *ACE*, **2**, 1235 (UV–visible).

S. Ahrland, *CIC*, **5**, 473 (UV–visible).
J. P. Hessler and W. T. Carnall, *LAS*, 349 (UV–visible).
N. M. Edelstein and J. Goffart, *ACE*, **2**, 1361 (magnetic).
J. W. Gonsalves *et al.*, *Inorg. Chim. Acta*, **21** (1977) 167; *S. Afr. J. Chem.*, **30** (1977) 62 (magnetic).

For the elements covered in sections 3.12–3.20, the following references should particularly be consulted: K. W. Bagnall, *CCC*, **3**, 1129; various authors, *CIC*, vol. 5.

Section 3.12

GMEL, Supplement No. 40, Main Volume 1942, Supplement 1981.
H. W. Kirby, *ACE*, **1**, 14.
C. Keller, *Chem. Zeitung*, **101** (1977) 500.

Section 3.13

GMEL, Supplement No. 44, Main Volume 1955, Supplements A1–A4 (elements), C1–C7 (compounds), D1–D2 (solution), E (coordination compounds).
L. I. Katzin and D. C. Sonnenberger, *ACE*, **1**, 41.
J. F. Smith *et al.*, *Thorium, Preparation and Properties*, Iowa State University Press, 1978.
L. Grainger, *Uranium and Thorium*, Newnes, 1958.
I. Santos *et al.*, *Adv. Inorg. Chem. Radiochem.*, **34** (1989) 65.

Section 3.14

GMEL, System No. 51, Main Volume 1942, 2 Supplements 1977.
D. Brown, *AIP*, 343; *Adv. Inorg. Chem. Radiochem.*, **12** (1969) 1.

Section 3.15

F. Weigel, *ACE*, **1**, 169.
E. H. P. Cordfunke, *The Chemistry of Uranium*, Elsevier, 1969.
W. Bacher and E. Jacob, *HAC*, **4**, 1 (UF_3).
J. Selbin and J. D. Ortego, *Chem. Revs*, **69** (1969) 657 (U(V)).
U. Castellato *et al.*, *Coord. Chem. Revs*, **36** (1981) 183 (nitrate complexes).
D. Brown, *MTP2*, **7**, 111 (nitrate complexes).
D. Brown, *MTP1*, **7**, 87 (halide complexes).
K. W. Bagnall, *MTP1*, **7**, 139 (thiocyanate complexes).
I. Santos *et al.*, *Adv. Inorg. Chem. Radiochem.*, **34** (1989) 65.
W. G. van der Sluys and A. P. Sattelberger, *Chem. Revs*, **90** (1990) 1027 (alkoxides).

GMEL, Supt. No. 55, Main Volume 1936, Supplements from 1975; A1–7 (element), B2 (alloys), C1–14 (compounds), D1–4 (solution), E1–2 (coordination compounds).

For the uranyl ion, see, for example:
S. P. McGlynn and J. K. Smith, *J. Mol. Spectrosc.*, **6** (1961) 164.
R. G. Denning *et al.*, *LAS*, 313; *Mol. Phys.*, **37** (1979) 1109.
W. R. Wadt, *J. Amer. Chem. Soc.*, **103** (1981) 6053.
P. Pyykko *et al.*, *Inorg. Chem.*, **28** (1989) 1801.

For the transuranium elements, *GMEL*, Supplement No. 71, 1973–79 should be consulted.

Section 3.16

J. A. Fahey, *ACE*, **1**, 443.
S. K. Patel, *Coord. Chem. Revs*, **25** (1978) 133.

Section 3.17

G. T. Seaborg, *AIP*, 1.
F. Weigel, J. J. Katz and G. T. Seaborg, *ACE*, **1**, 499.
J. M. Cleveland, *The Chemistry of Plutonium*, American Nuclear Society, 1979.
W. T. Carnall and G. R. Choppin (eds), *ACS Symp. Series* 1983, vol. 216.

Section 3.18

W. W. Schulz and R. A. Pennemann, *ACE*, **2**, 887.
J. D. Navratil, W. W. Schulz and G. T. Seaborg, *J. Chem. Educ.*, **67** (1990) 15.
N. M. Edelstein, J. D. Navratil and W. W. Schulz, *Americium and Curium Chemistry and Technology*, Reidel, 1985.

For syntheses of trans-americium halides see, for example, J. R. Peterson *et al.*, *J. Inorg. Nucl. Chem.* **35** (1973) 1525, 1711; **40** (1978) 811; **43** (1981) 2425; *J. Radioanal. Chem.*, **43** (1978) 479; *Inorg. Chem.*, **25** (1986) 3779.

Section 3.21

T. J. Marks, *HAC*, **4**, 491; *ACE*, **2**, 1588.
T. J. Marks and A. Streitweiser, *ACE*, **2**, 1547.
T. J. Marks and I. L. Fragala (eds), *Fundamental and Technological Aspects of Organo f-element Chemistry*, Reidel, 1985.
T. J. Marks and R. D. Ernst, *Comprehensive Organometallic Chemistry*, Pergamon, **3** (1982) 173.

Section 3.22

E. K. Hulet *et al.*, *J. Inorg. Nucl. Chem.*, **42** (1980) 79.
R. J. Silva, *ACE*, **2**, 1103.
I. Zvara, *Inorg. Nucl. Chem. Lett.*, **7** (1971) 1107.
A. Ghiorso, *AIP*, 23.
G. T. Seaborg and O. L. Keller, Jr, *ACE*, **2**, 1629.
Y. T. Organessian *et al.*, *Radiochim. Acta*, **37** (1984) 113.
P. Armbruster *et al.*, *Ann. Rev. Nucl. Part. Sci.*, **35** (1985) 135; *J. Less Common Metals*, **122** (1986) 581.
B. Fricke, *Structure and Bonding*, **21** (1975) 89 (predicted properties).

For an outline of nuclear stability, see:
G. T. Seaborg, *J. Chem. Educ.*, **46** (1969) 626.

Index

NOTE: owing to the general similarity of lanthanide(III) compounds, the following simplified procedure has been adopted. A comprehensive index is found under the heading LANTHANIDES. A compound such as, say, praesodymium(III) chloride is only indexed under PRAESODYMIUM if specifically mentioned in the text. General information relevant to PrCl will however be found by looking up 'chlorides' under the LANTHANIDES heading.